# HOW THE BRAIN
# EVOLVED LANGUAGE

· · · ·

# HOW THE BRAIN
# EVOLVED LANGUAGE

· · · ·

Donald Loritz

# OXFORD
UNIVERSITY PRESS

Oxford   New York
Athens   Auckland   Bangkok   Bogotá   Buenos Aires   Cape Town
Chennai   Dar es Salaam   Delhi   Florence   Hong Kong   Istanbul   Karachi
Kolkata   Kuala Lumpur   Madrid   Melbourne   Mexico City   Mumbai   Nairobi
Paris   São Paulo   Shanghai   Singapore   Taipei   Tokyo   Toronto   Warsaw

and associated companies in
Berlin   Ibadan

Copyright © 1999 by Donald Loritz

Published in 1999 by Oxford University Press, Inc.
198 Madison Avenue, New York, New York 10016

First issued as an Oxford University Press paperback, 2002

Oxford is a registered trademark of Oxford University Press

All rights reserved. No part of this publication may be reproduced,
stored in a retrieval system, or transmitted, in any form or by any means,
electronic, mechanical, photocopying, recording or otherwise,
without the prior permission of Oxford University Press.

Library of Congress Cataloging-in-Publication Data
Loritz, Donald, 1947–
How the brain evolved language / Donald Loritz.
p.  cm.
Includes bibliographical references and index.
ISBN 0-19-511874-X; 0-19-515124-0 (pbk.)
1. Language and languages—Origin.  2. Biolinguistics.  3. Grammar,
Comparative and general.  4. Human evolution.  I. Title.
P116.L67   1999
401—dc21      98-29414

3 5 7 9 8 6 4

Printed in the United States of America
on acid-free paper

*This book is dedicated to my family
and
to the memory of
Walter A. Cook, S.J.*

# ACKNOWLEDGMENTS

This book has been so long in the writing that I can hardly begin to recount and thank the many people who have helped it on its way. Foremost among them are certainly Paula Menyuk and Bruce Fraser, who introduced me to the many, many subtleties of generative psycholinguistics. Then I must thank Steve Grossberg for teaching me to avoid "homunculi" as cheap solutions to hard psychological problems.

In the late 1970s and 1980s, there was little appreciation of neural networks. During these years I must thank in particular Jan Robbins, Walter Cook, and Sam Sara for supporting me on little more than simple faith. By the late 1980s, however, neural networks had become more widely respected, and with the help of Bernard Comrie, Charles Ferguson, Winfred Lehmann, and Betty Wallace Robinett, and the support of my colleagues, I succeeded in securing tenure and setting aside some time to think heretically about how language could possibly work without a computational homunculus.

For such sense as the following pages might make, I am indebted also to Allen Alderman, Kathy Broussard, Craig Chaudron, Dick Chen, Brad Cupp, Lisa Harper, Beatriz Juacaba, Jee Eun Kim, Elena Koutsomitopoulou, Bernard Kripkee, Mark Lewellen, Lise Menn, Afsar Parhizgar, Lisette Ramirez, Bill Rose, and Elaine Shea for their insights and support. I fear I have not yet achieved the rigorous standards my teachers have set, but the time has come to collect my attempts at explaining language so that better minds might bring them to fruition.

*December 1998*                                                              D. L.

# CONTENTS

ONE · Lought and Thanguage   3

TWO · Jones's Theory of Evolution   21

THREE · The Communicating Cell   36

FOUR · The Society of Brain   52

FIVE · Adaptive Resonance   74

SIX · Speech and Hearing   90

SEVEN · Speech Perception   109

EIGHT · One, Two, Three   123

NINE · Romiet and Juleo   133

TEN · Null Movement   143

ELEVEN · Truth and Consequences   161

TWELVE · What If Language Is Learned by Brain Cells?   171

*Notes   195*

*References   203*

*Index   219*

# HOW THE BRAIN
# EVOLVED LANGUAGE

• O N E •

# Lought and Thanguage

A baby wildebeest, born on the Serengeti, learns to walk and run in a matter of minutes. The precocious wildebeest is marvelously adapted. But once a wildebeest can outrun a lion, its adaptation is largely done. Humans must adapt, too, but they must continue to adapt, first to mother, then to brother, then to teacher, then to work, then to him, then to her, then to babies, and so it goes. The world to which humans must adapt is a world of words: his word, her word, my word, your word, word of mouth, word of honor, word of scripture, word of law, the adaptive word.

A baby wildebeest, born on the Serengeti, learns to walk and run in a matter of minutes, but it takes two full years before a baby human makes a word, learns to talk. And it is another ten years before the human child learns to talk back. Even then, it is an immature criterion that would claim the teenager has mastered language. Human beings' fascination with language persists as long as their fascination with life. How does a baby learn language? How does one talk to a teenager—or, for that matter, to a parent? What do words mean? Or, since *I* know what *I* mean, what do *your* words mean? And what do *you* think *my* words mean? Does she think I'm a nerd? Does he think I'm a bimbo? Will my boss think I'm disloyal if I don't say *yes*? How can I adapt?

## Where Is Language?

One doesn't have to be a philosopher to ask such questions, and in fact, it may be better if one is not. The celebrated learned men of history never concerned themselves much about the learning of language. (After all, teaching children language is women's work!) Thus absolved of responsibility for explaining language in its most human terms, the wise men of history were freed to conflate and confuse thought and language as they pleased.

This is not to deny that down through the ages many amusing books have been written on the topic of language. For example, in 1962 the philosopher J. L. Austin published a book exploring the question *Is the King of France bald?* (Since there is no longer a King of France, Austin concluded the question is neither true nor false, but simply *void.*) By and large however, previous philosophers of thought and language have simply been ignorant of one important modern fact: *thought happens in the brain!*

Aristotle, to pick one particularly influential example, thought that the main function of the brain was to cool the blood.[1] In hindsight, the ancients' ignorance of the brain and its function was quite understandable. Locked up in its bony carapace, the brain, which resisted exposure to the warrior's sword, resisted as well the anatomist's scalpel. And even when the ancients noticed there was a brain beneath all that bone, they couldn't see it *do* anything. It didn't beat or breathe or bend. What ancient could have imagined that the brain created ideas with the same electrical forces as Zeus's thunderbolts? Real knowledge, Aristotle thought, was lodged in the *heart,* and even today, when we know something well, we say we "know it by heart." So we can understand as well the ancients' belief that knowledge was mysteriously dissolved in the blood.

Finally, 2,000 years after Aristotle, Harvey showed that the heart pumped blood through the body, circulating nourishment to the organs of the body. Knowledge had to be somewhere else. But the microscope had not yet been invented, and when the seventeenth-century eye looked at the brain, the first feature it noted was that the brain, like the heart, had several connected, fluid-filled chambers, called *ventricles.* (In Figure 1.1, a *horizontal section* of the brain exposes the main, lateral ventricles.) To seventeenth-century philosophers, the meaning was obvious: the brain was just another pump. Following Galen and Harvey, Descartes thought it pumped an animating fluid (*animus*) through the nerves, thereby causing muscles to move. He specifically thought that the pineal gland at the base of the ventricles was a kind of master valve, which controlled hydraulic pressure in the system. To Descartes, this brain-pump was just so much more plumbing. Hydraulically moving the muscles was important, but it was just machinery; it could have nothing very much to do with thought. For Descartes, thought happened somewhere else. Thought happened in the *mind,* not in the brain.

But where was the mind? For Descartes, language, mind, and thought were all essentially the same thing. Descartes would have asserted that it makes no more sense to ask *Where is the mind?* than it does to ask *Where is language?* or *Where is algebra?* Such questions, to use Austin's term, were simply void. Language, thought, and mind were abstract sets of formal relations. They could relate things in places to other things in other places, but they were not themselves in some place. For Descartes, thought and language, mind and meaning, algebra and geometry, were all essentially the same sort of thing, which is to say they weren't really *things* at all.

In the seventeenth century this dualism of mechanics and mind, of things-in-the-world and things-not-in-the-world, had a confirming parallel in the Church's natural-supernatural dualism of life and afterlife. In a sense, Descartes

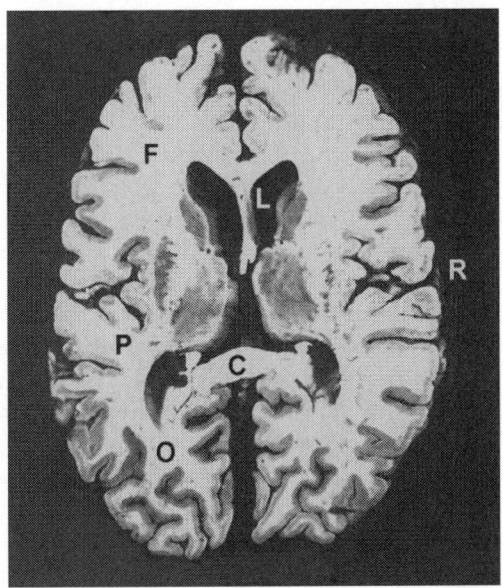

**Figure 1.1.** Horizontal section of the cerebrum. The (lateral) ventricles are exposed. Occipital lobe (O), parietal lobe (P), frontal lobe (F), lateral ventricle (L), corpus callosum (C), fissure of Rolando (R). (Kirkwood 1995, 15. Reprinted by permission of Churchill Livingstone.)

extended the conception of the supernatural to include not only angels but also algebra, algorithms, and language. These otherworldly entities had a truth that, to Descartes, was obviously true, a priori, before and independent of any empirical experience. One could only find this truth by doubting empirical, in-the-world experience and by believing in a priori, not-in-the-world truths, truths like the existence of God.

But how could you or I, mere mortals both, know in the mind that even we ourselves exist, let alone so sublime a being as God? Was there a less presumptuous a priori truth from which we could deduce these larger Truths? Perhaps the most famous "rationalist" deduction of this sort was Descartes's proof of his own existence:

*cogito ergo sum* (1.1)

think-I therefore exist-I

"I think therefore I am."

Unfortunately, as Nietzsche later observed, Descartes's "proof" turns out to be uselessly circular: in Latin, the *-o* on *cogito* and the form *sum* itself both indicate first-person *I*. Consequently, as the literal gloss of 1.1 emphasizes, the premise *I think* presupposes the conclusion *I am*. To illustrate this point, consider 1.2 and 1.3 (here and elsewhere a * means "something seems wrong with this sentence"):

*You think therefore I am. (1.2)

or

*Thought exists therefore I am. (1.3)

Without its presuppositions, Descartes's proof fails utterly. In Descartes's defense, we should perhaps consider the context of his times. The Reformation had put reason at odds with God, and Descartes had a larger agenda than to vainly prove his own existence. But the proof is still false. Even a genius cannot deduce truth from faulty premises.

## Tabula Rasa

Well before Nietzsche, many philosophers objected to Descartes's dualistic method. Descartes's contemporary Francis Bacon strenuously objected to Descartes's introspective method. Francis Bacon (and, coincidentally, 400 years earlier, Roger Bacon) espoused a rather distinctively English empiricism. Unlike Descartes's dualism, this empiricism was a triadism that divided the universe into Soul, Mind, and Matter. Leaving the supernatural aspects of Soul to God, empiricism proceeded to focus on the material aspects of Matter. But neither Bacon was a rigorous scientist by modern standards. (In what was apparently his first and only scientific experiment, Francis Bacon stuffed a chicken with snow to see if the snow would inhibit decay. The only reported result was that Bacon caught cold and died.) The relationship of Mind to Soul and Matter was little advanced by their methods. It wasn't until a hundred years after Descartes that empiricism found a clear voice in the philosophy of John Locke. For Locke, Mind was just a blank slate, an erased tablet of Matter, a tabula rasa. Experience wrote upon the tablet, thus creating Mind. Of course rationalists objected that this explained no more than *cogito ergo sum*. If, as the empiricists would have it, there was such a tablet, then where was it? Where was Mind? And if this tablet were writ upon in language, then where was language? Void questions all! So rationalism survived until 1849, when Claude Minié invented the conical bullet.

Before 1849, bullets were musket balls. Musket balls had a frustrating habit of curving unpredictably in flight, so prior to 1849, opposing armies would line themselves up, shoulder to shoulder, in order to give the opposing team a reasonable chance. Even then, when a musket ball did happen to score, it tended to shatter the skull, causing massive damage to the brain beneath. Minié's conical bullet, on the other hand, flew true. Even better, it was frequently able to create a surgically clean hole in the skull and a nice, focused wound (a *focal lesion*) in the underlying brain tissue.

As a result of this technological advance, a young doctor in France, Pierre Paul Broca, obtained a sizable cohort of war casualties whose brain lesions disturbed their ability to speak but otherwise left the patients' minds and behaviors remarkably intact. In 1861, Broca presented the discovery that such *aphasia* occurred almost exclusively when injury was sustained to a relatively limited area of the left half of the brain. Several years later a Viennese doctor, Karl Wernicke, discovered that injuries to another region on the left side of the brain caused a second kind of aphasia. Whereas "Broca's aphasics" had difficulty speaking but relatively little difficulty comprehending language, "Wernicke's aphasics" had no difficulty speaking but great difficulty comprehending. *Where is language?* had seemed a void question, but suddenly—and quite unexpectedly—it had an answer.

## Where Language Is

Language was in the brain! This finding, utterly implausible to the ancients, was supported by copious and irrefutable evidence: spoken output was generated in Broca's area, and heard input was processed in Wernicke's area. The scientific community instantly and earnestly undertook the study of the brain.

It was no longer the seventeenth century. Leeuwenhoek had long since invented the microscope, and within a generation of Broca, scientists had trained it on the brain. In 1873, Camillo Golgi discovered that chromium-silver salts would selectively stain brain cells, thus making them clearly visible under the microscope. Using Golgi's staining method, Santiago Ramón y Cajal charted the microstructure of the brain in encyclopedic detail, and by the dawn of the twentieth century, it had become an established scientific fact that mind was brain. And since brain was made up of white matter and gray matter, mind was matter. Rationalism was dead.

For their discovery of the brain's previously invisible structure, Golgi and Ramón y Cajal were awarded the 1906 Nobel Prize.[2] Their work also engaged them in a famous debate. Ramón y Cajal believed each cell was a separate cell, wholly bounded by its cell membrane and unconnected to its neighbors, but his microscopes weren't powerful enough to prove it. On the other hand, Galvani had long before shown that electricity made a dissected frog's leg twitch. It could therefore be readily inferred that there was electrical communication among nerve cells. But how could electrical impulses be transmitted if the wires weren't connected? Golgi maintained that the myriad cells of the nervous system must form a continuous network.

In the early 1900s many more researchers joined in this debate. Using ever-more-powerful microscopes, they took ever-closer looks at nerve cells. In the end, Sherrington, Adrian, Dale, Loewi, and others proved that Ramón y Cajal was right, earning in the process Nobel Prizes for their efforts. Neurons were discrete cells separated by a synaptic gap. This gap was small, but it was big enough to electrically insulate each cell from the next. So how did neurons pass their messages across the synapse? They passed their electric messages using chemicals, called *neurotransmitters*. Doubt of the world and belief in truth were now clearly behaviors of the brain:

> Thus, both doubt and belief have positive effects upon us, though very different ones. Belief does not make us act at once, but puts us into such a condition that we shall behave in a certain way, when the occasion arises. Doubt has not the least effect of this sort, but stimulates us to action until it is destroyed. This reminds us of the irritation of a nerve and the reflex action produced thereby; while for the analogue of belief, in the nervous system, we must look to what are called nervous associations—for example, to that habit of the nerves in consequence of which the smell of a peach will make the mouth water. (Peirce 1877:9)

## Never Mind the Mind

But Peirce was ahead of his time. Twenty years later, in America, Peirce's "pragmatic" perspective developed into behaviorism. Behaviorism came in many flavors, but one lineage descended from Peirce to Dewey to Thorndike to Watson to Lashley. In the formulation of John B. Watson, behavior could be observed and scientifically reduced to a series of *stimulus-response* events, "habits of the nerves," occurring along a chain of neurons. Mind was just an unobservable and useless abstraction. All of creation, from the lowliest animal to the highest form of social organization (then widely believed to be either the assembly line or the Prussian army), could be pragmatically analyzed solely in terms of stimulus-response chains of command. Behaviorism, in the social form of totalitarianism, promised a well-regulated society in which every animal want could be provided by eliciting strict, learned, obedient responses to the stimuli of an all-powerful, all-loving dictator.

Predictably, this utopian vision was especially popular among the ruling and managerial classes, who had never worked on an assembly line or directly experienced the new, improved, conical bullet. Many, following Herbert Spencer (1862) and later "social Darwinists," envisioned themselves to be "supermen," a new species which had evolved through natural selection to a point "beyond good and evil" (Nietzsche 1883). However, after World War II and the likes of Hitler and Stalin, this utopian vision began to lose some of its appeal, even among the controlling classes. In his 1948 utopian novel *Walden Two*, the celebrated Harvard behaviorist B. F. Skinner attempted to dissociate behaviorism from these infamous European practitioners. As Skinner spun the story, everyone—more or less regardless of race, creed, color, or, for that matter, genetics—could be educated to perfection through the application of "programmed learning." In programmed learning, students were methodically rewarded for correct answers and punished for incorrect answers. In this way, it was believed that good habits would be efficiently "learned" and bad habits would be efficiently "extinguished."

In the United States, however, there was a new class of university students: World War II veterans whose college tuition was paid as a war benefit. These students and vocal, war-hero labor union leaders let it be known that they did not consider *any* chain of command to be utopian. Whether on the front line, the assembly line, or the school registration line, they did not want to be programmed! By the mid-1950s, opposition to Skinner had become widespread, but it was inchoate. Behaviorism had been politically refuted by the European experiment with totalitarianism, but Skinner's scientific authority as a Harvard professor was still unassailable, and there were no viable alternatives to his psychological theories.

In 1957, amid mounting popular disdain for behaviorism, Skinner published a scholarly book, *Verbal Behavior*. In it, he sought to show that behaviorism had developed far enough beyond the study of lab rat behavior to undertake the explanation of human language. In 1959, two years after the publication of *Verbal Behavior*, Noam Chomsky, a young linguist at the Massachusetts Institute of Technology, published a disdainful review of it in the journal *Language*. Not only did Chomsky find Skinner's analyses of language naïve, but he found them to be proof of the vacuity of behaviorism in general.

Skinner didn't reply directly to Chomsky's review, but he did write another book, *Beyond Freedom and Dignity*, to which Chomsky also gave a bad review. These reviews of Skinner and behaviorism made Chomsky an instant, popular champion of freedom and dignity, opening a new chapter in the confusion of thought and language.

## Finite Mind, Infinite Language

Reaching back to rationalism for support, the thrust of Chomsky's argument was that language was not a "thing" like a stimulus or a response, a punishment or a reward. Language was a unique—and uniquely human—module of mind. Thus, twentieth-century *generative grammar* became grafted onto a Cartesian dualism. The resulting generative philosophy has depended heavily on what I call the "generative deduction," the basic form of which may be given as follows:

(1a) The human brain is finite, but
(1b) an infinity of sentences exists,
(1c) which can be generated by rule,

proving language is infinite. Nevertheless,

(2a) normal human children acquire language quickly and effortlessly,
(2b) even though no one teaches language to young children,
(2c) and only human children so acquire language.

Therefore,

(3) language is innate. It is not so much learned as it is "acquired."

The premises of the generative deduction have come under attack from many quarters, but it has not yet been refuted. Consider, for example, Jackendoff's 1994 witty defense of premise 1. First, Jackendoff opens the dictionary at random and generates a large number of sentences by a simple rule:

> A numeral is not a numbskull.
> A numbskull is not a nun.
> A nun is not a nunnery.
>
> . . .
>
> These are all completely absurd, but they are sentences of English nevertheless. There will be something like $10^4 \times 10^4$ of them = $10^8$. Now let's put pairs of these sentences together with *since*, like this:
>
> . . .
>
> Since a numeral is not a numbskull, a numbskull is not a nun.
>
> . . .
>
> Since a numeral is not a numbskull, a numbskull is not a nunnery.
>
> . . .
>
> And so on it goes, giving us $10^8 \times 10^8 = 10^{16}$ absolutely ridiculous sentences. Given that there are on the order of ten billion ($10^{10}$) neurons in the entire human brain, this divides out to $10^6$ sentences per neuron. Thus it would be impossible for us to store them all in our brains. (Jackendoff, 1994:21)

Although $10^{16}$ does not quite qualify as mathematical infinity, it certainly seems infinite for human purposes. This infinity of language was at the nub of Chomsky's arguments against Skinner in 1959, and premise 1 of the generative deduction has stood unrefuted and irrefutable until the present day.

For the past forty years, a variety of biologists, psychologists, teachers, and child-language researchers have contested premise 2, arguing that children *are taught* language and do in fact *learn* in the process. But premise 1 forms the basis for a strong logical defense of premise 2. Chomsky has introduced that defense with a different quotation from Peirce:

> You cannot seriously think that every little chicken that is hatched has to rummage through all possible theories until it lights upon the good idea of picking up something and eating it. On the contrary, you think that the chicken has an innate idea of doing this; that is to say, that it can think of this, but has no faculty of thinking anything else. . . . But if you are going to think every poor chicken endowed with an innate tendency towards a positive truth, why should you think to man alone this gift is denied? (Peirce, quoted in Chomsky 1972, 92)

Peirce called the ability to come up with new theories *abduction,* a logico-cognitive process which he believed was more important than either of the logical processes of induction or deduction. Chomsky asked essentially the same question of children and language: one cannot seriously think every little child that is born has to rummage through all possible grammatical theories until it lights upon the one right way of making words into sentences. Language could not be learned unless every child was endowed with an innate tendency toward a correct, universal grammar.

Following Chomsky's suggestions, researchers undertook a series of mathematical analyses, collectively referred to as "learnability theory," to investigate the conditions under which language could be learnable (Gold 1965, 1967; Hamburger and Wexler 1975; Wexler and Culicover 1980; see Pinker 1984, 1989, for approachable reviews). The gist of their argument was the following. If you say *potayto* and I say *potahto*, how is a child to learn which one to say? This argument becomes more convincing as one considers, not just the 5,000 or 10,000 words that a child might memorize, but also the fact that the child knows how to transform these words à la Jackendoff into an infinite number of sentences (premise 1 again). Chomsky's seminal example was the "passive transformation," as of 1.4 into 1.5:

John saw her. (1.4)

She was seen by John. (1.5)

Instead of 1.5, why doesn't a child ever say 1.6*,

*Saw by John was she. (1.6)

or any of the other 118+ possible permutations of 1.5? "Because the child never hears those other 118+ permutations," you may say. But the child has likely never heard the exact permutation which is 1.5, either. Nevertheless, every child has learned to produce passive sentences like 1.5 by the age of six or so (premise 2a).

"Well, the child doesn't memorize rote sentences," you reply. "He remembers patterns." But exactly how *does* he remember patterns? No one in his right mind sits down and teaches a child of four that "to transform an active sentence pattern into a passive sentence pattern, one positionally exchanges the subject and direct object, prefaces the subject with the word *by*, appropriately changes the grammatical case of the moved subject and direct object, precedes the main verb with the tensed auxiliary of *be*, agreeing in number and person with the new subject, and replaces the main verb by its past participle."

You might instead argue that the child learns language patterns by *imitating* adult speech, and this was in fact the explanation proposed by behaviorists. Unfortunately, child-language researchers quickly found that children don't imitate adult speech. Consider the following, oft-quoted transcript from McNeill 1966:

Child   Nobody don't like me.
Mother   No, say "Nobody likes me."
Child   Nobody don't like me.
Mother   No, say "Nobody likes me."

Child   Nobody don't like me.
Mother  No, say "Nobody likes me."
Child   Nobody don't like me.
Mother  No, say "Nobody likes me."
Child   Nobody don't like me.
Mother  No, say "Nobody likes me."
Child   Nobody don't like me.
Mother  No, say "Nobody likes me."
Child   Nobody don't like me.
Mother  No, say "Nobody likes me."
Child   Nobody don't like me.
Mother  Now listen carefully. Say "Nobody likes me."
Child   Oh, nobody don't likes me.

To maintain that language is "learned," it appears one needs a better theory of learning than imitation.

Although generative philosophy has demonstrated the failure of behaviorism to most observers, it has not been without its critics. For example, the claim that language is rule-based (premise 1c) extends back to the foundations of modern linguistics in the eighteenth century, but for forty years, nonlinguists have objected that language cannot be rule-governed, because *semantics*, the meaning system of language, is not rule-governed. After all, what rule could definitively tell you what I mean when I say *I love you*? But semantics has little to do with the generative deduction. Chomsky has argued that "such understanding as we have of [language] does not seem to be enhanced by invoking thoughts that we grasp, public pronunciations or meanings, common languages that we partially know, or a relation of reference between words and things" (1993, 25), and as Jackendoff's *A nun is not a nunnery* illustrates, sentences can be grammatical even if they are meaningless. That is, leaving meaning aside, how is one even to explain *syntax*, if not as acquired through the agency of an innate, rule-governed system?

Recently, many cognitive psychologists have attacked premise 1c by demonstrating that pattern-based neural networks can exhibit linguistic behaviors similar to that of rule-based systems (Rumelhart and McClelland 1986a). But to date these demonstrations have been more semantic than syntactic. Also, the fact that rulelike behavior can be elicited from an artificial neural network does not preclude the possibility that the brain functions at some other, more interesting level like a rule-based digital computer.

My discomfort with the generative deduction originated with premise 2a, that children learn language "effortlessly." To be sure, childhood in middle-class America in the latter half of the twentieth century has been mostly child's play, but even privileged children display the temper tantrums of the "terrible twos," and these are nothing so much as results of the child's frequently frustrated *efforts* at communication. Nor do mommies and au pairs find the terrible twos "effortless." Nevertheless, the claim that toddlers learn language effortlessly seems never to have been challenged directly, and I am unaware that generative philosophers have ever independently proposed an objective measure of child effort. The problem, no doubt, is that *effort* is an intrinsically subjective, "nonscientific" concept. Society devalues child labor because no one pays children a salary, and no one hears children complain—no one except mommies and au pairs, that is, but "scientific" society doesn't pay *them* salaries, either.

"Hard science" often tries to distance itself from such social issues, but when the object of scientific inquiry is language, it is hard to maintain distance. As a kind of compromise, sociolinguists (Ferguson and Slobin 1973) and "functionalists" (Bates and MacWhinney 1982; MacWhinney 1987a) have attacked premise 2b by redefining learning away from the narrow terms of behaviorism into more general terms of interaction in the social environment. We learn that the sky is blue, that birds fly, and that ice is slippery from the physical environment without a teacher, but no one claims *this* knowledge is innate. Sociolinguistic functionalism argues that we learn language from the social environment in much the same way. But how *do* we learn that birds fly and ice is slippery? Generative philosophers have justifiably objected that this sort of learning (*a*) is not itself well understood and so (*b*) barely begins to address deeper problems like how we understand the sentence *I don't think penguins can fly*.

Finally, biologists have often attacked premise 2c, the human uniqueness of language, citing dancing bees and signing apes as evidence of the evolution and learning of language in other species. Nevertheless, not even the proudest trainer invites his animals to cocktail parties. Whatever their language, animals' language is still a far cry from human language.

Although locally convincing, none of these attacks has proved generally fatal to the generative deduction, much less added up to a viable alternative theory of thought or language. Taken together, though, they indicate that something is amiss with the generative deduction. Forty years after first postulating that children have an innate "language acquisition device," generative philosophers have as yet been unable to find its place in human biology, and generative theory has found itself increasingly at odds with the rest of science and society. Chomsky himself has become defensive, asserting that "no one knows anything about the brain" (Chomsky 1988, 755), and asking,

> how can a system such as human language arise in the mind/brain, or for that matter, in the organic world, in which one seems not to find systems with anything like the basic properties of human language? That problem has sometimes been posed as a crisis for the cognitive sciences. The concerns are appropriate, but their locus is misplaced; they are a problem for biology and the brain sciences, which, as currently understood, do not provide any basis for what appear to be fairly well-established conclusions about language. (Chomsky 1994, 1)

The preceding is neither a crisis for biology nor a crisis for linguistics; it is a crisis for Science. The assertion that no one knows anything about the brain may have been defensible in 1936, when Turing initiated "the study of cognitive activity from an abstract point of view, divorced in principle from both biological and phenomenological foundations" (Pylyshyn 1979). It may also have been defensible in the late 1950s, when the foundations of generative philosophy were being laid. But since then, some two dozen Nobel Prizes have been awarded for discoveries in brain science. To date, at the end of the twentieth century, some thirty-three Nobel Prizes have been awarded for discoveries about the human brain and nervous system (Ramón y Cajal in 1906, Golgi in 1906, Sherrington in 1932, Adrian in 1932, Dale in 1936, Loewi in 1936, Erlanger in 1944, Gasser in 1944, Hess in 1949, Békésy in 1961, Hodgkin in 1963, Huxley in 1963, Eccles in 1963, Hartline in 1967, Wald in 1967, Granit in 1967, Axelrod in 1970, von Euler in 1970, Katz in 1970, Guillemin in 1977, Schally in 1977, Yalow in 1977, Sperry in 1981, Hubel in 1981, Wiesel in 1981, Levi-Montalcini in 1986, Cohen in 1986, Sakmann in 1991, Neher in 1991, Fischer in 1992, Krebs in 1992, Gilman in 1994, Rodbell in 1994). The problem today is not that "no one knows anything about the brain." The problem is that we know so much about the brain and its abnormalities in so much detail that it becomes difficult to step back and see how the brain might do something so normal and so large as language.

## The Von Neumann Limit (v)

The great mathematician John Von Neumann is credited with having invented the modern serial computer's organization into "procedural memory" (program) and "declarative memory" (data). Although Ramón y Cajal had won the 1906 Nobel Prize for showing that the brain is a massively parallel processor, Von Neumann declared that "the nervous system [is] a computing machine in the proper sense, and that a discussion of the brain in terms of the concepts familiar in computing machinery is in order" (Von Neumann 1958, 75).

Von Neumann went on to claim that since serial computers could do everything parallel computers could do and then some, they were in principle of design superior to parallel computers. It follows logically from this premise that the digital computer is, in principle of cognitive design, superior to the human brain, and that the computer scientist could be in this respect superior to God Almighty. Of course, most computer scientists have been too modest to make these deductions in public, but in the privacy of classified documents, these obvious implications sold a lot of computers to a world military/security establishment bent on being almighty (Roszak 1986). Flush with money and power from these contracts, Von Neumann and his followers overlooked one small factor in their calculations, however, death.

It turns out that the fatal flaw in the generative deduction is in its least-examined premise, premise 1a. While a serial computer might well be able to do everything a parallel computer can do, it can't always do those things in the $10^9$-odd seconds of a human's allotted lifetime. The reason every little child doesn't have to rummage through all of the words and sentences his little head can hold is not so much that every little child is born with a language acquisition device endowing him with an innate tendency toward universal grammar.

It is that *the human mind is infinite*. To see how this is so, let us recall Jackendoff's conclusion:

> And so on it goes, giving us $10^8 \times 10^8 = 10^{16}$ absolutely ridiculous sentences. Given that there are on the order of ten billion ($10^{10}$) neurons in the entire human brain, this divides out to $10^6$ sentences per neuron. Thus it would be impossible for us to store them all in our brains.

Recall how Jackendoff got to these numbers. He combinatorially paired $10^4$ words into $10^8$ simple sentences and then combinatorially paired those $10^8$ simple sentences to create $10^{16}$ compound sentences. But if words and sentences can combine, *why can't neurons combine?* In fact, Ramón y Cajal showed that neurons *do* combine: each brain cell makes synaptic connections to thousands and thousands of other brain cells. Brain cells function in subnetwork combinations with other brain cells. And just how many subnetwork combinations can brain cells make, you ask? Well, it so happens there is a formula for *combinations:*

$$\frac{n!}{(n-k)!k!} \qquad (1.7)$$

If we assume that each of Jackendoff's $10^{10}$ neurons is used in language and that each makes some $10^3$ synaptic connections, and if we further assume that every word is represented by $10^6$ connections combining in simultaneous activation, then with $n = 10^{13}$ synapses taken in combinations of $k = 10^6$, 1.7 becomes 1.8:

$$v = \frac{10^{13}!}{(10^{13} - 10^6)!10^6!} \approx 10^{7,111,111} \qquad (1.8)$$

There is no computer large enough to compute the value of v exactly, but $10^{7,111,111}$ is a reasonable approximation.[3] And how big is $10^{7,111,111}$? Well, $10^{300}$ is a generous estimate of the number of atomic particles in the known universe, so not only can your brain store Jackendoff's $10^{16}$+ sentences, it can also store a name for every particle in the universe! And it will still have room for naming all the particles in $10^{7,110,811}$ more universes!! Compared to Jackendoff's $10^{16}$ sentences—or the brief candle of a human life—your mind's capacity is, for all human purposes, infinite.

But are we to seriously think that every little child that is born has to rummage through all possible $10^{7,111,111}$ combinations until it lights upon the one good grammatical idea for a sentence? Only if we process them serially. Because letter follows letter, word follows word, and sentence follows sentence, the wise men of history from Aristotle to Von Neumann supposed that serial language must be the product of a serial process: because language is serial, they supposed thought must also be serial.[4] But unlike Von Neumann's serial computer, it takes no more time for a parallel-processing brain to abduce $10^{7,111,111}$ theories than it does to abduce 1 theory! Think about it this way. When you go to the zoo and see a zebra, do you start going through all the names in your head—*aardvark, ant, anteater, antelope* . . . —until you come to *zebra?* Of course not. *Zebra* comes to mind as quickly when you see a zebra as *aardvark* does when you see an aardvark. Neither is this trick of fast, "content-addressable" memory unique to humans. When a lion sees an aardvark or a zebra, I'm sure it knows immediately what's for dinner, even though it doesn't have a printed menu. When a pigeon sees a hawk or a fox, it immediately decides whether to duck or to fly. In the meantime, the little serial computer, which had to rummage through all its possible plans of action, would have been lunch. If we are prepared to think that every poor pigeon is innately endowed with a fast, content-addressable, parallel-processing brain, why should we think to *Homo loquens* alone this gift is denied?

## Adaptive Grammar and the Plan of the Book

It is important not to confuse thought and language. Just because language is manifestly serial, it does not necessarily follow that language must be computed by a serial processor. Although language is serial, thought is parallel. Although Turing machines are serial processors, the human brain is a parallel processor.

While generative philosophy's fundamental confusion of thought and language cannot be ignored, the primary objective of this book is not an attack upon generative linguistics or artificial intelligence. Within their just premises, generative linguistics has identified numerous previously unnoticed cognitive phenomena which demand explanation, and serial computers have often proved themselves to be valuable, if not very adaptive, devices. On the other hand, children are not Turing machines, and so long as children are needed, they need a theory of their own. What is needed is a theory of how human language, which functions to serve our minute-by-minute social adaptation, arose as part of the same adaptive, evolutionary process which led to *Homo loquens*. This book attempts to develop such a theory. Chapters 2–5 outline the theory's foundations, from elementary evolutionary and biological principles of life through Stephen Grossberg's adaptive resonance theory (ART). Chapters 6–12 apply ART to language, creating in the process that specific application of ART which, to give my critics a convenient target, I will call *adaptive grammar*.

Adaptation occurs on many timescales. Over the ages, each phylum evolved as an adaptation to a changing Earth. Over the millennia each species evolved as an adaptation to changing habitats. These are the timescales on which adaptation is usually discussed. But each individual human also must adapt to society over the course of a lifetime, and the neurons which encode our daily thought and language must adapt even faster, on a scale of seconds. English speakers, for example, must adapt every *a* or *the* to their listeners' knowledge. Thus, we say *a dog* until our listener knows which dog we're talking about, and thereafter, we say *the dog*. On these latter timescales, it is common usage to call adaptation *learning*.

Because adaptation is such a universal property of life, adaptive grammar is rooted in many disciplines, and this book must range from Cambrian to Cenozoic, from molecules to minds to language, across evolutionary biology, neurobiology, psychology, linguistics, mathematics, and computer science. Even if I could master all of these fields and their subfields, I would need to simplify, and even still, every reader would find some chapter difficult and another chapter oversimplified. For this, and for the many lacunae in my own knowledge, I beg the reader's indulgence. In the end, however, a viable adaptive grammar will be the product of many minds, and to attract a broad resonance, I have sacrificed much detail to readability.

Chapter 2 begins by taking a long, long view of adaptation and adaptive communication, starting with the evolution of the brain cell out of the "primordial soup." That soup has left few fossils, so this is an admittedly speculative view, but it is justified on three accounts. First, the story of evolution establishes the biological principle of *self-similarity*, which is needed to extend adaptive grammar beyond the simplest biological and linguistic examples. Second, imagining how the simplest, two- and four-celled brains evolved is a preliminary thought experiment in Grossberg's method of minimal anatomies, the reasonable method of explaining complex neural processes in terms of simpler neural processes. This preliminary thought experiment shows how basic neural mechanisms that would eventually be needed for language had already evolved as early as 600 million years ago. Finally, the story of evolution coincidentally gives the uninitiated reader a narrative thread that can tie together the numerous biological facts upon which subsequent chapters will build.

Chapter 3 studies the single-celled organisms called neurons. But whereas chapter 2 was necessarily speculative, there is relatively little that is speculative about chapter 3. Twentieth-century science has laid neurons out plainly before us, photographed by electron microscopes and dissected with biochemical scalpels. With the rapid progress of biological science, every speculation in this domain quickly becomes a testable—and tested—hypothesis.

The neuron is ultimately a social creature, so chapter 4 brings the neuron's billion-year evolution to its culmination in the complex society of the human brain. Chapter 4 first looks at the large-scale organization of the brain as it appeared to Broca and Wernicke: hindbrain, midbrain, and forebrain, right hemisphere and left hemisphere, front and back. These views are now fairly familiar to the educated reader, but chapter 4 at its end takes some more unusual, less traditional views: it also looks at the brain top to bottom, inside out, and splayed flat.

Once upon a time, there were four myopic neuroscientists. Peering through the microscope, they happened upon a brain cell. The first, observing a cell body shaped like a pyramid, said, "This is a pyramidal cell." The second, observing the spines on the cell's dendrites, said, "This is a spiny cell." The third, observing its spherical neurotransmitter vesicles, said, "This is a spherical cell." The fourth, noting the concentration of glutamate in those vesicles, said, "This is an excitatory cell." Of course, they all were right, but by using ever-thicker lenses to study ever-smaller objects, our myopic neuroscientists eventually "could no longer see the forest for the trees." Only a handful of neural network researchers persisted in studying the neural forest that Ramón y Cajal discovered, explored, and charted.

Following these researchers, chapter 5 turns off the electron microscope and uses Grossberg's adaptive resonance theory to describe the midscale organization of the brain. Earlier theories, like behaviorism, were built on the details of how individual neurons behaved (as described in chapter 3), while later theories, like generative philosophy, were built on the organization of gross modules of the brain like Broca's area and Wernicke's area (described in chapter 4). But these theories could not integrate their levels of description to explain how an entire brain full of neurons is capable of writing *War and Peace*—or, for that matter, guiding me to my office every morning—even though no individual neuron supervises the process. ART, by contrast, describes such a brain in terms of minimal anatomies that are organized on a scale of two to a few thousand neurons. Only when we understand how and why these minimal anatomies work do we begin to understand how and why larger systems like thought and language must work as they do.

Adaptive resonance theory has been constructed principally by the analysis of mathematical models. Insofar as adaptive grammar is more concerned with its linguistic validity than its mathematical validity, chapter 5 will only touch upon the mathematical foundations of ART. To help the general reader visualize many of the most important features of ART—contrast enhancement, noise suppression, resonance, self-similarity, and neural rebounds—without close mathematical study, the main points of chapter 5 are presented by means of a graphical computer simulation. This will be a relief to many readers, but it would be wrong to suppress the mathematics entirely. Unlike first-year calculus's thin gruel of missile trajectories, the mathematics of ART is humane and exciting, so I have tried to keep just enough of it to encourage the intrepid reader to venture into ART's primary literature.

Linguistics often views language as atomic sounds (*phones*) built up into the progressively larger structures of phonemes, morphemes, words, phrases, sentences, and discourse. It happens that this approach also suits the method of minimal anatomies, so chapters 6–12 follow the same general plan, applying ART to these successive levels of language, in the process constructing that corpus of explanations I call adaptive grammar.

Chapter 6 is transitional between neuroanatomy and linguistics. It describes how the mouth produces phones and how they are received by the ear and passed along auditory nerve pathways to the brain.

After the physical, physiological, and phonetic description of speech and speech sounds in chapter 6, the problem of *phonemes* is addressed in chapter 7. This is the basic problem of how you can say *potayto* and I can say *potahto*, yet we can both still mean the same tuber. Similar questions have been investigated by ART, but these have mostly been about vision—for example, how we can stably identify grass as "green" even though daylight itself changes in color from dawn to noon to dusk. At this phonemic level, where neurons in the minimal anatomies between the ear and Wernicke's area process the sound spectrum of speech, previous ART analyses map quite directly onto issues of linguistic perception.

But speech spectra change more quickly than Apollo drives his team across the sky: language happens quickly in time, and serial order becomes a fundamental problem. For example, when children first learn to count, they tend to count "one, two, three, eight, nine, ten." What happens to the middle of this series? Behaviorist accounts in terms of stimulus-response chains (*one* stimulates *two* stimulates *three*, etc.) obviously had some missing links, but the issue is also critical for parallel models of cognition: how *can* a parallel brain encode serial order? Among major "connectionist" theories, only ART offers a sufficiently detailed analysis of neural architecture to solve this fundamental linguistic problem. Chapter 8 outlines the general solution.

Intending to honor the queen, Rev. William Archibald Spooner, Fellow and Warden of New College, Oxford, is claimed to have offered a toast to "our queer, old dean." In his memory, such quaintly twisted phrases now bear the slightly derisive eponym of *spoonerisms*. Psychologists, however, are not at all derisive. In 1951, it was the lowly spoonerism that spelled the beginning of the end for behaviorism. In that year, Karl Lashley (a student of J. B. Watson, the "Father of Behaviorism" himself) first noted the impossible problem spoonerisms posed for behaviorism: in spoonerisms, stimulus-response chains not only lose some links but they must also split apart and rejoin in totally nonhabitual recombinations. Worse, Lashley noted that these recombinative reversals occurred frequently and ubiquitously not only in disorders like dyslexia but also in normal speech and common behaviors like dance and tpying [*sic*]. Chapter 9 explains the spoonerism as a natural interaction of ART rebounds (chapter 5) and ART serial-learning anatomies (chapter 8). Rather unexpectedly, this leads to a deeply rhythmic analysis of word structure (morphology), one uncannily reminiscent of recent linguistic theories of "metrical phonology."

Lashley died in 1958, so it was left to Chomsky to administer the coup de grace to behaviorism. Chomsky saw that spoonerisms were not only isolated error phenomena but actually instances of a more general linguistic process called *metathesis*. In particular, Chomsky saw that sentences like 1.9 and 1.10 could also be related by metathesis:

Spoonerisms slew behaviorism. (1.9)

Behaviorism was slain by spoonerisms. (1.10)

Chomsky's theory of linguistic metathesis was built on the mathematical principles of Alonzo Church's *lambda calculus* (1941). The lambda calculus is a recursive grammar of algebra and the design specification for a kind of computer called a *pushdown-store automaton*. Pushdown-store automata are especially suited to recursive operations upon data that are structured in *binary trees*, so Chomsky explained that sentence 1.10 was derived from 1.9 by a "passive transformation" on syntactic trees, which moved *behaviorism* to the tree position of *spoonerism* and vice versa. Chomsky's theories generated widespread and well-funded enthusiasm: if recursion and pushdown-store automata could explain language, then certainly they could explain human intelligence in general! Although Chomsky cautiously distanced himself from such glib enthusiasm, his work became very much a cornerstone of a generation's research in "artificial intelligence."

But if language is produced by brain cells, what could it mean to "move *behaviorism* to the tree position of *spoonerism*"? Could it mean that in transforming 1.9 into 1.10, some *behaviorism*-cell and some *spoonerism*-cell actually exchange places in the brain? Of course not. So chapter 10, "Null Movement," states the obvious and rejects Chomsky's basic explanation of metathesis: nothing moves. It then extends the analyses of chapters 7–9 to develop an alternative explanation of how words can "move" if neurons can't. In place of movement, chapter 10 borrows the serial organization principles of chapter 8 and organizes syntax around the neural representation of *topicality*.

Chapter 11 is a kind of cadenza. As we progress from phoneme to morpheme to phrase to sentence, we require larger and larger minimal anatomies to describe phenomena. When we finally reach the stage of social discourse and meaning, our anatomies are no longer minimal, and as we approach the Von Neumann limit, they take on the behavior of free will. Just as Einstein rejected Heisenberg's uncertainty principle,[5] many "hard" scientists today reject the notion of free will, but when, as sometimes happens, the activity of 1 subnetwork in $10^{7,111,111}$ causes an entire brain to change its "mind-set," scientific prediction and explanation can no longer be 100% accurate. In chapter 11 the multiple themes of the earlier chapters become intertwined in ways that science is reluctant to entertain. Truth and meaning are explored as logical and social constructs, respectively, and self-similarity is reinvoked with the epigram "A human being is a neuron's way of making another neuron."

Chapter 12, "What If Language Is Learned by Brain Cells?" is like ontogeny recapitulating phylogeny. The themes of the preceding chapters are recapitulated, but this time from the perspective of the individual language learner. At the outset, the unborn fetus is not only affected by its genetic inheritance but also exposed to a raft of environmental hazards, so phenomena of disordered language learning are treated first. When I began this book after many years away from the language disorders literature, I confess to having been skeptical of the many and varied complaints of "learning disability." But as it evolved, adaptive grammar itself began to convince me not only that learning disabilities were real but also that nearly everyone is learning disabled. But just as the chapter appears ready to end with a prescription of learning pills for everyone, it returns to a reconsideration of learning—normal language learning, nature's way of enabling a less-than-perfect assemblage of neurons in the human brain to adapt and survive. Our computers are wonderful and our medicines are wonderful, but it is most wonderful of all that nearly every human child, despite genetic defects and a hostile environment, learns a human language and survives.

• T W O •

# Jones's Theory of Evolution

> The Sanskrit language, whatever be its antiquity, is of a wonderful structure; more perfect than the Greek, more copious than the Latin, and more exquisitely refined than either, yet bearing to both of them a strong affinity, both in the roots of verbs and in the forms of grammar, than could possibly have been produced by accident; so strong, indeed, that no philologer could examine them all three, without believing them to have sprung from some common source, which, perhaps, no longer exists.
>
> Sir William Jones (1786)

We identify transformational grammar with Chomskyan grammar, but ideas about language "transformations" were in the air well before Chomsky's name was attached to them. We call the heliocentric solar system the Copernican system, but Aristarchus first proposed it nearly 2,000 years before Copernicus. We call the theory of evolution "Darwin's theory of evolution," but the central notion of evolution was proposed nearly 100 years before Darwin by the eminent English philologist Sir William Jones. Following Jones, European philologists embarked upon the reconstruction of what they called the "Aryan language," the ancestor of all modern Indo-European languages. (Because Hitler misappropriated the term "Aryan," this ancestral language is now called "Proto-Indo-European.") Within a generation, it had been conclusively proved that languages as diverse as English and Sanskrit had, in fact, descended from this now-extinct language. Since these philological reconstructions did not extend back more than a few thousand years, they did not challenge the biblical account of creation, and Jones's theory of evolution quickly became established scientific fact. Within a generation, the philologists' ideas and methods were adopted by Lamarck, Chambers, Wells, and many other predecessors of Darwin who proposed the evolution of nonhuman species within biblical time. So Darwin did not really so much invent the theory of evolution as apply it correctly to biology.

But even after the philologists explained the descent of modern languages and Darwin explained the descent of man, there remained something about humankind that evolutionary theories seemed still unable to explain. That something was descent of language—not just English or French or Chinese, not even Proto-Indo-European, but language itself. This great web of meaning that we humans call language—can it not be called our soul, or at least our mind? And however might evolution account for that? For centuries, fabulous fakirs and sober scientists had attempted to produce horses that count and dogs that talk. In no case, however, did any succeed in producing or finding animal language that seemed anything like human language. Absent Darwinian evidence that language has evolved gradually, generative philosophy set forth the bold assertion that language is a specific and uniquely human development, undescended from Darwinian nature. Generative philosophy allowed this development to be physiologically associated with brain regions like Broca's area and Wernicke's area, but it implied that, in Chomsky's words, any attempt to study language as having evolved from general animal intelligence was "adaptationist hogwash" (Chomsky 1988). This further implied not only that behaviorism had been defeated by language but that its essential procedure, the extrapolation of research findings about animal behavior to human behavior, was sterile.

Beginning in the 1960s, generative psycholinguists therefore began to explore language anew, as an autonomous module of mind. But by the mid-1970s, there were some disturbing reports from these frontiers. Researchers into sign languages of the congenitally deaf (especially ASL, American Sign Language; Klima and Bellugi 1979) had established that these visual languages were capable of expressing human thought as completely as spoken language. We will revisit this topic in chapter 9, but since sign language does not involve primary speech cortex or hearing cortex, language began to look like an adaptation of a more general intelligence.[1]

Then Philip Lieberman showed that apes and other quadrupeds had been incapable of speech, not because of some profound cognitive deficit, but simply because of the shape of their vocal tracts. As we shall see in chapter 6, speech sounds depend upon the shape of the vocal tract. When the human species assumed an erect posture, the vocal tract bent 90°, its acoustic properties changed dramatically, and speech became possible. But apes' vocal tracts remained unbent and physically incapable of producing most speech sounds—despite the best efforts of fabulous fakirs and sober scientists.

Armed with these two pieces of evidence, ethologists stopped trying to teach apes to speak and started trying to teach them ASL. A series of chimps and gorillas (Gardner and Gardner 1969; Patterson 1978; see Premack 1985 for a summary) rapidly learned vocabularies of hundreds of signs, and what is more, they learned to combine them into new signs and sentences. For example, upon seeing a swan, a chimp called Washoe combined the signs for *water* and *bird.*

Thus, an evolutionary link was established between animal language and human language—at least in the minds of many biologists and ethologists. But generative philosophers remained unimpressed. To them, Washoe had signed nothing more remarkable than "there's some water; there's a bird." Apes still had not demonstrated the ability to generate novel and infinite sentences. Syntax remained a summit which only humans had scaled.

Nevertheless, by the turn of the twenty-first century, even some of his MIT colleagues began to abandon Chomsky's hard line (Pinker and Bloom 1990, Pinker 1994). It became acceptable, even in some linguistic circles, to say that language had evolved. The naked ape stood upright, bending his vocal tract. This gave him a significantly enriched inventory of "calls," and the rest is history. Unfortunately, this account still does not explain some of the more remarkable aspects of language, let alone of mind. The generative objection is that this account does not explain why, for example, we can say sentences 2.1 and 2.2 but not 2.3:

The man who is$_1$ dancing is$_2$ singing a song. (2.1)

Is$_2$ the man who is$_1$ dancing singing a song? (2.2)

*Is$_1$ the man who dancing is$_2$ singing a song? (2.3)

For a deeper explanation of language, we must look deeper into evolution. We must go back in time, long before the hominids, long before the prehominids. In order to understand how human language and communication make survival possible, we must understand how intercellular communication among the first one-celled life forms made multicelled life forms possible. And to understand this, we must go back long before even the dinosaurs, back to when the only living things were rocks.

Consider that a rock lives and dies. Take a common crystal of baking soda (sodium bicarbonate, $NaHCO_3$), and drop it into a supersaturated solution of baking soda. Behold, the crystal grows. And while it is true that there is more to life than growth, note how this growth is like life. The growth of a crystal is self-similar: a small crystal grows into a large crystal, and just as a small boy grows into a big boy, its essential structure remains unchanged.

Moreover, crystals reproduce. If we split our crystal in two, each half grows into a crystal which is also self-similar to its parent. And while it is true that there is more to life than growth and reproduction, consider also that our poor crystal can die.

Baking soda is hardly a diamond among crystals, but it is spectacular in its death. Complete this elementary-school thought experiment by dropping a teaspoon of vinegar (flavored acetic acid) on our crystal. Immediately, the crystal undergoes a fiery death, leaving behind only a pile of soda ash, a puddle of water, and a cloud of carbon dioxide. Thus, even the lowliest rocks have the essentials for a tale of life, death, and, sometimes, transfiguration. What is wanting is a little more personality, a salt of the earth with a little more spice.

In 1951, looking for that spice, Stanley Miller concocted a primordial soup of water, methane, ammonia, and hydrogen. In his laboratory, he subjected this mixture to artificial lightning, seeking to simulate early conditions on the planet Earth. After a week, Miller began to find amino acids in his soup. At the time, only a handful of scientists appreciated what Miller had found, but in 1953, Watson and Crick suggested "a structure for the salt of deoxyribonucleic acid" (DNA). This salt of DNA organized the assembly of amino acids into proteins, the stuff of life.[2]

One molecule of baking soda is just like any other molecule of baking soda. What gives proteins and DNA personality is their complexity. DNA chains four nucleic acid bases—guanine (G), cytosine (C), adenine (A), and thymine (T)—along a backbone of phosphate and deoxyribose-sugar molecules. Each subchain of three bases defines a *codon*, which in turn defines one of twenty amino acids. Chains of codons form genes, and chains of amino acids form proteins.

DNA is also sexy. One chain of nucleotides does not make a complete DNA molecule; it must be embraced by a second chain, its complementary image. When DNA reproduces, the two chains unwrap, and each builds a new mirror-image partner (figure 2.1).

It is still a long way from an amino acid to a sexy, self-replicating strand of DNA, and although Miller's experiment has been superseded by much more sophisticated studies (e.g., Orgel 1979; Cairns-Smith 1985; see Dawkins 1986 and Dennett 1995 for highly readable introductions to evolutionary biology), his experiment is still superb allegory, for Stanley Miller's test tube did not only bring forth amino acids; it also brought forth that old villain, acetic acid. Life, death, and simple salts. The life of the first nucleic acids cannot have been much less treacherous than the life of a baking-soda crystal in a third-grade classroom. Even the earliest molecules of life had to be constantly on the lookout for killer chemicals.

Despairing of the odds of bringing forth life in a hostile primordial soup (even given two billion years), I did not know how to continue this chapter. Then I gazed out my window at the rain, and I saw a test tube in every drop and a laboratory in every puddle, and when I multiplied billions of drops by billions of puddles by billions of years, the emergence of complex self-replicating molecules no longer seemed so miraculous. Like a Polynesian explorer happening upon an island without snakes and without disease, all that life needed was one puddle of Eden, protected against predators by an impermeable barrier of rock.

*Nearly* impermeable, that is, because at some point, on some day, some one self-replicating, complex molecule had to bravely go where no such molecule had gone before, and it had to survive. To survive outside paradise, it had to clothe itself; it had to bring a barrier with it. In billions of puddles over billions of years, probably billions of colonies of replicating molecules tried on many different clothes, but in a watery world, it was the *cell* that survived.

## The Cell

In the cell, a bilayer of phospholipid hydrocarbon molecules created a puddle within the primordial puddle. A phospholipid hydrocarbon is a fatty molecule, one end of which is electrochemically repelled by water while the other is at-

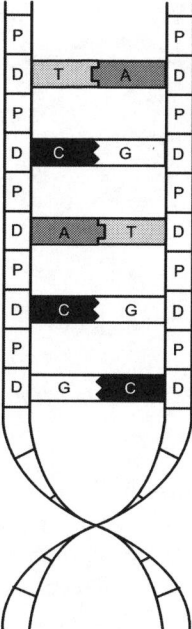

**Figure 2.1.** DNA.

tracted to it. A bilayer of such molecules forms the cell's membrane, a molecular, water-repellent balloon enclosing the complex and vulnerable self-replicating DNA molecules and protecting them against chemicals that would do them harm. But just as the primordial puddle's rocky barrier had to be slightly permeable to let the first intrepid self-replicating molecule out into the world, the growing and surviving cell design also needed a semipermeable membrane, so that nutrients could enter and waste products could be expelled. And this posed a significant evolutionary dilemma: how was the membrane to let food molecules in but keep toxic molecules out?

Nature needed three basic solutions to this problem. The first was the communal solution, and it is still visible to every third grader who examines a drop of pond water under the microscope. Like little walled cities, the *eukaryotic* cell (figure 2.2) housed within its membrane mitochondria, ribosomes, and various organelles: cooks, builders, and housekeepers that process incoming food, build new cells and cell parts, and keep things clean and healthy.

It seems that early cells simply ingested one another, sometimes being the feeder and sometimes being the feed. But on a few very special occasions, symbiotic relationships formed. Mitochondria, ribosomes, and other organelles all learned to live together in self-similar domestic harmony: membranes within membranes, and within each membrane, molecular economies importing raw materials and exporting products, communicating with each other in chemical codes. Primus inter pares, at the heart of every eukaryotic cell, is the cell

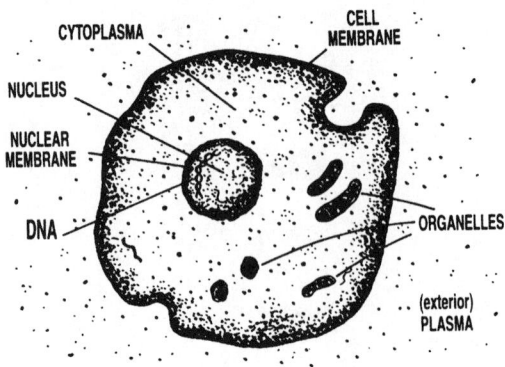

**Figure 2.2.** The eukaryotic cell.

nucleus. The cell nucleus is an organelle, but it is a very special organelle. It is the organelle that contains DNA, the plans by which the biological city rebuilds itself. Now, toxins and other bad guys who scaled the cell's first, outer cell wall could kill it, but in order to kill the eukaryotic cell's *progeny*, toxins also had to penetrate the second, nucleic membrane.

The second solution, in important respects self-similar to the first but especially important to the eventual evolution of nerve cells, was the evolution of the bilayer membrane itself: one layer keeps the outside out while the other keeps the inside in. The outside membrane has "gates" and "latches" that are highly nutrient-specific and only open when a nutrient is identified. The inside membrane has gates that open when wastes are present, allowing them to be expelled.

One can imagine that the first cell membrane was selective but passive, allowing molecules to penetrate it only by osmosis. Like the shoreline of the primordial pond, which admits some runoff from puddles on higher ground and trickles some of its contents to ponds on lower ground, such an arrangement would allow select molecules at high concentration to pass to regions of lower concentration. This passive membrane could recognize nutrient and waste molecules by their shape so as to ingest only the former and excrete only the latter, but Precambrian life was not so civilized that a protozoan epicure could float around until a molecular morsel was served with perfect presentation. The price of a too-passive and too-refined selectivity could too easily be starvation. The surviving and *thriving* protozoan needed to be a very picky eater only until it found something it liked. Then it needed to pig out, which leads us to the third solution to the problem of selective permeability.

Fortunately, molecules become ionized in solution: they assume an electrical charge. In the primordial saltwater soup, sodium chloride dissolved into positively charged sodium ions ($Na^+$) and negatively charged chloride ions ($Cl^-$). As a result, when an early protozoan encountered a field of nutrients, it could do more than passively wait for them to seep through its membrane. So long as it kept an internal negative charge, when it opened one of its mouths (i.e., one of the pores of its membrane) to eat a big, tasty protein molecule, a flood of the ubiquitous $Na^+$ ions from the surrounding seawater soup would be electrically attracted into the cell interior as well. The resulting change in voltage would literally *shock* the surrounding membrane, causing two more mouths to open, admitting more $Na^+$, creating a still bigger shock, opening still more mouths. Like screaming "Pizza!" in a crowded college dormitory, the influx of $Na^+$ made the membrane active, setting off a feeding frenzy. As we will see in chapter 3, this active membrane is a hallmark of the *neuron*, the cell type which sends electric signals to other cells. The other hallmark of the neuron is its *axon*: a kind of long transmission cable along which its electric signals are sent.

## Flagellates

It is all well and good for a cell to sit in its little pond, swallowing anything that floats its way with an appealing shape and turning up its pores at the rest. But to find more food, grow big, and eat all the other little cells, it helps if a cell can move around. One of the first groups of moving cells, the *flagellates* subphylum (*Mastigophora*), is particularly intriguing. To move around, these single-celled animals evolved a long, whiplike protoplasmic projection called a *flagellum* (figure 2.3). The flagellum would also have been an excellent prototype for an axon, so the flagellates would have been an excellent prototype for the neuron. But a neuron with an axon to send an electric nerve signal is still missing one thing: a cell to receive its message. Perhaps it is not wholly accidental that the flagellates formed colonies, possibly making the important evolutionary transition to *Porifera* (sponges) and *Coelenterata* (jellyfish and coral), the first multicelled animals.

At this still-early stage of evolution, multicellular organization is often selfsimilar to cellular organization. In primitive colonial plants like slime mold and colonial protozoans like Porifera, when a cell divides, the two new cells do not swim apart; they stick together. Mobility is sacrificed for a different advantage. The cells on the outside of the colony have greater access to food but also risk greater exposure to toxins. Inner members have less access to food but are less

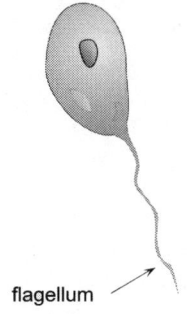

flagellum

**Figure 2.3.** A flagellate.

exposed to toxins, and should outer cells die, the inner cells can still live long enough to reproduce.

In the case of differentiated multicellular organisms like jellyfish, the organism's outer membrane differentiates first into outside (the skin, or *epithelium*) and inside (the gut, or *enteron*). Then the cells of the animal differentiate further into such types as muscles for movement and nerves for control and communication.

## The Cambrian Explosion

Up until about 600 million years ago, life left few fossils. Presumably, most life up to that time was one-celled and soft—not the stuff of which fossils are made. But then, quite suddenly in the geological record, at the boundary between the Permian and Cambrian periods, fossils of multicelled animals begin to appear in great profusion and variety. What could have caused this *Cambrian explosion* of life? One factor certainly must have been the invention of the neuron, for, by definition, the multiple cells of a multicellular organism must communicate among themselves in order to function as an organism. While the electrical communication of the neuron might not have been strictly necessary (the organs of your body also communicate by hormones sent through the bloodstream), the race in life is between the quick and the dead, and there is little doubt that the quick animals of the Cambrian were electric-quick: they had electric neurons. Evolution was no longer just a process, it was a race—an evolutionary arms race (Dawkins and Krebs 1979), and the neuron was its conical bullet.

## The Formula for Life

The reader may not be ready to agree that Miller's creation of amino acids was the same thing as creating life. Indeed, recent discoveries of thermophilic life forms (the *Archaea*, which live off sulfur, deep in the sea) have made the events of early evolution seem bizarre indeed, and we now know that Earth's early environment couldn't have been quite the combination Miller concocted. Still, even if we haven't found the specific formula for life, we can be quite sure we have found the general formula. It is stated in equation 2.4:

$$f(x) = Ax + t \qquad (2.4)$$

Of course, if $A$, $x$, and $t$ in equation 2.4 are simply real numbers, then 2.4 is nothing but a straight-line function. But if $x$ is a form and $A$ is a geometrical transform of that form, then 2.4 is the formula for an *affine transformation*, more popularly known as a *fractal*. In the case of figure 2.4, $x$ is the almost-triangular shape of a fern leaf; $A$ is a not-very-complex function which rescales and tilts the shape of $x$; and $t$ simply relocates the transformed shape to a different place in the picture. The result is that each frond of the fern in figure 2.4 is another whole frond, but on a different scale in a different place. Following Mandelbrot 1982, this property of the "fractal geometry of nature" is often called *self-similarity*, a term we have seen again and again, the principle that patterns in nature repeat themselves on different scales, albeit with slight mutations.

I don't believe, as some mystics might, that equation 2.4 implies that there is either a mathematical design or a mathematical precision behind Creation. Equation 2.4 only simply and succinctly captures a basic pattern of nature, $x$. And because it only produces the fern of figure 2.4 when parameters of $A$ and $t$ are procedurally varied, it also focuses our attention on essential processes of nature. Life is one such process, which faces similar problems over and over again, albeit in different places ($t$) and on different scales ($A$). Life first prospered in a puddle, within a barrier. Similarly, on another scale, the life of the cell occurs within a barrier, the cell membrane; and on yet another scale, the life of the cell nucleus occurs within another membrane barrier. On yet other scales, each organ of your body is surrounded by a membrane barrier; your body itself is surrounded by a membrane barrier called skin. Cities are surrounded by walls, nations by borders, and Earth by an atmosphere. You may find mystery in this if you like, but this is also just how things are. In all the preceding examples, despite variations in detail, there is a common, elemental force which imposes this design on all scales of life. We could say that this force is the second law of thermodynamics, that membranes are barriers against entropy. We could say that epithelia are barriers against predators. We could say that barriers establish the identity of Self versus Other. All of the above could be true, and more, but what is essential to our present story is that the barrier design works to enable self-replication. What works may not be True, but what works Survives.

**Figure 2.4.** A fractal fern. (Barnsley 1988. Reprinted by permission of Academic Press.)

## Sex and Self-Similarity

Amid all this self-similar evolution there is also differentiation. In colonial Protozoa and Porifera, any member of the colony can, in theory, split off and begin a new colony, self-similar to its previous colony. That means its DNA must contain the plans to the whole city. But by the time we move up the evolutionary ladder to Coelenterata, things get more complex.

In jellyfish we begin to see a very clear differentiation of cells: there are muscle cells and stinger cells and brain cells. Given such a differentiated jellyfish, how is a jellyfish to make another jellyfish? In a protozoan, say a shapeless amoeba, it is simple in principle. Let it simply divide in half, and voila! one has two amoebae. But how do you divide a salamander in half? A salamander can grow a new tail, but a tail can't grow a new salamander. How is even a jellyfish to make a new jellyfish? In multicellular organisms, cells become specialized, and so sperm cells and egg cells become specialized for reproduction. The basic idea is pretty obvious: keep some eggs around. What is really astonishing is that every jellyfish (not to mention every human!) must be rebuilt from scratch, from a single set of DNA plans contained in a fertilized egg. And there is basically only one way to do this: the same way nature did it. The one-celled egg cell must evolve into a multicelled organism with a skin and a gut, an inside and an outside, your basic coelenterate. Then the basic coelenterate, if it hopes to be a human, must evolve a backbone, making the evolving human embryo your basic fish. Then, if this basic fish is to become a human, it must develop limbs, making your basic fish look like your basic reptile. After all, one could hardly develop fingers before arms.

The individual evolves in a fashion similar to the way his species evolved. In biology, this principle is captured in the phrase *Ontogeny* (the development of the individual) *recapitulates phylogeny* (the development of the species within a phylum). We will return to this principle in chapter 12, but for the present, it is pertinent to see that this is not only just another manifestation of self-similarity, but a *procedural* self-similarity. Like the structural self-similarity I have been describing, it is simply the path of least resistance through a series of evolutionary problems. How can nature keep toxins from the crystals? First build a membrane. How can nature build fingers and toes? First build limbs. This principle is classically illustrated in figure 2.5, in which fowl, rabbit, and human embryos all begin by looking alike (and rather fishlike) but then diverge as they develop.

But who needs sex? Puritans think we would all be better off without it, and a few plants and animals do seem to do nicely without it. The key word here seems to be "few." In the evolutionary arms race, species must seek a new and better offense (or a new and better defense). This means species must *change*. One can accomplish this change by sitting around in the primordial puddle, waiting for an act of God to effect a mutation, or one can go out and actively swap DNA, combinatorially accelerating change in the fractal equation of life. While only a few species reproduce asexually, the myriad species inhabiting every nook and cranny of Earth's ecosystem are the product of sex.

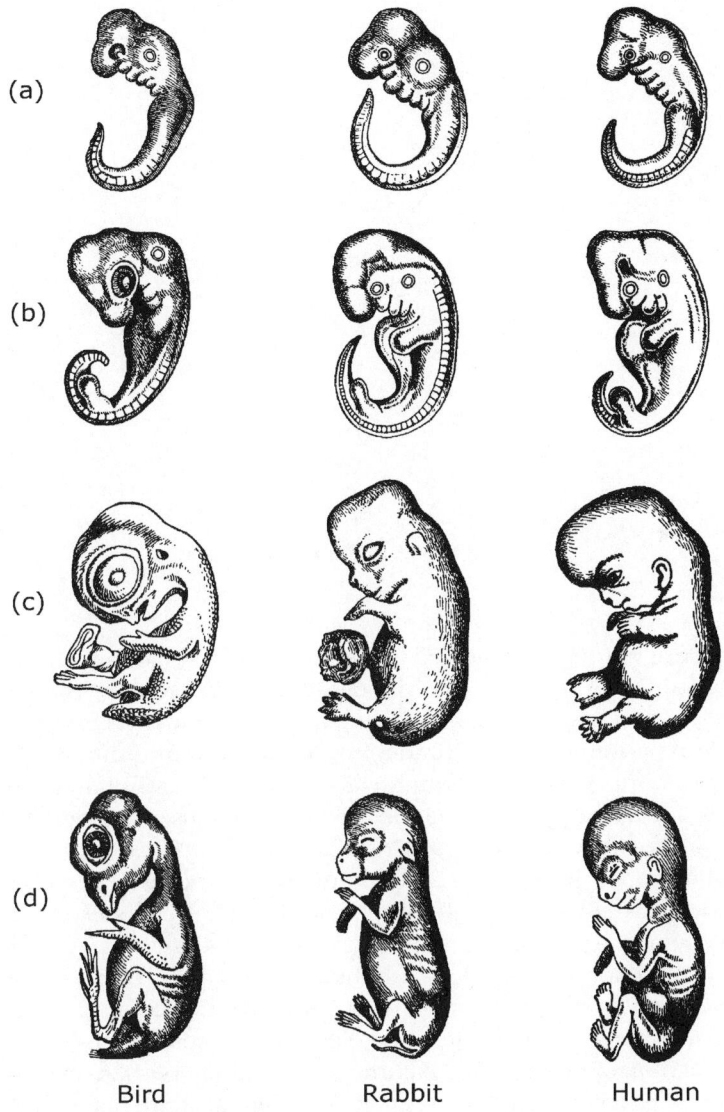

**Figure 2.5.** Ontogeny recapitulates phylogeny. (von Baer 1828. Reprinted by permission of the New York Public Library.)

## Some Basic Brains Evolve: The One-Celled Brain

Just as the evolutionary criterion of survival dictated how simple organisms had to be structured and how they had to reproduce, there seem to exist only a few paths along which the modern neuron could evolve and survive. One path is exemplified by arthropods like the crayfish. In the crayfish and other lower phyla, the nervous system takes on the appearance of a serial anatomy (figure 2.6).[3] In

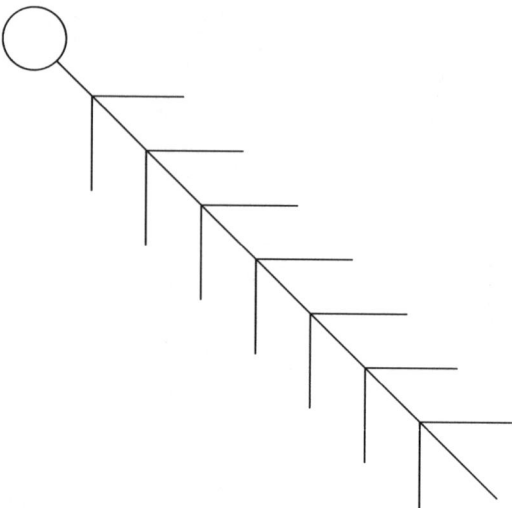

**Figure 2.6.** An avalanche anatomy.

this brain design, sometimes called an "avalanche," a single signal, originating in the crayfish brain, is sent down the tail. Each swimmeret of the tail is then successively innervated by an axon collateral, and the crayfish is propelled along, swimmeret by swimmeret. This branching of the axon into multiple axon collaterals is an important and ubiquitous feature of axons. The neuron doesn't really use its own energy to send its electrical signal; it uses the local $Na^+$, which surrounds its axon everywhere. As a result, it hardly takes any more energy to send a signal along a hundred or a thousand axon branches than it does to send a signal along one axon. As a result, even the lowly crayfish can use a single, central nervous system to coordinate a multitude of distal swimmerets.

Figure 2.6 is not intended to insult the intelligence of your average lobster; arthropods actually have many more than one cell in their brains. Rather, figure 2.6 is a *minimal anatomy*. It uses just one diagrammed cell to self-similarly represent a population of many cells. (Technically, we should perhaps call each "cell" or "neuron" of such a minimal anatomy by a more abstract name like "site" or "neurode" or "population." But concrete terms like "neuron" are more readable, and now that I have made my point, belaboring the reader with self-similar example upon self-similar example of self-similarity, I will strive for readability.)

A one-celled brain can be simple-minded in more than one way. Consider, by contrast with the crayfish, how the common jellyfish moves. To move, its brain gives a single command which contracts the enteron, forcing water out the rear and propelling the jellyfish forward. In order for this to work, all the muscles around the periphery must contract and relax together. This is accomplished with a radially branching axon, as in figure 2.7a. This minimal anatomy is only a slight evolutionary variation on figure 2.6: figures 2.6 and 2.7a are both "one-celled brains" with axon collaterals. But they are also significantly differ-

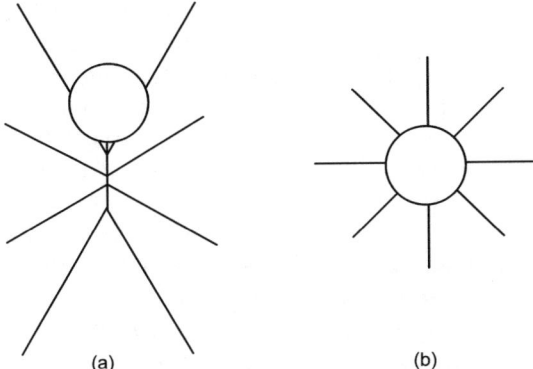

**Figure 2.7.** A radial anatomy, or "outstar."

ent in how they move and behave. Whereas figure 2.6 is a "serial anatomy," figure 2.7a is a "parallel anatomy": all terminal nodes are activated at once, in parallel. In figure 2.7b, this radial anatomy is further schematized as an "outstar." Note that such outstar minimal anatomies do not violate the biological fact that every cell has one and only one axon leaving its cell body. Outstar minimal anatomies are drawn to emphasize the parallel branching of axon collaterals or the many axons emanating from a population of many neurons.

## The Two-Celled Brain

The first two-celled brain may be supposed to have evolved when two neurons accidentally synapsed with each other. The result is the minimal anatomy of figure 2.8. Note first that this minimal anatomy implies a parallel organization.

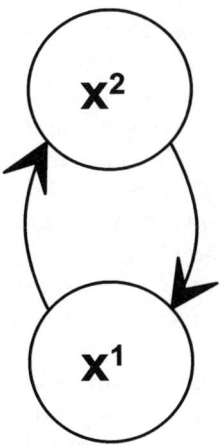

**Figure 2.8.** A two-celled brain.

No matter whether cell $x^1$ initially excites cell $x^2$ or vice versa, in short order they are both excited together, at once, in parallel. But like many other evolutionary accidents, the two-celled brain entailed both good news and bad news.

The good news is the following. Let either cell start "firing." It thereby activates the other cell, which thereby activates the first cell, and so on. We will say that the two cells *resonate*, and in chapter 5, adaptive resonance theory will show how this can be a very good thing. It is a form of memory, albeit only short-term memory. The bad news is that this resonance is also the model of a nervous system out of control, a kind of miniature model of an epileptic convulsion, and as in a convulsion, resonance and contraction cannot go on forever. Such a brain works for jellyfish: parallel activation causes the jellyfish's radial musculature to contract convulsively, and the jellyfish "swims" forward, but only when the convulsion stops from nervous exhaustion can the muscles relax and the jellyfish ready itself for another convulsive forward stroke. When coelenterates evolved, it was an evolutionary marvel that they could move at all, but in the race between the quick and the dead, other organisms soon found a faster way to move.

## The Six-Celled Brains

Several phyla evolved a six-celled brain model, most famously the phylum Chordata, the vertebrates, you and I. Like many other phyla, the chordates abandoned radial symmetry and evolved bilateral symmetry: their bodies have a left and a right side.

Unlike most surviving phyla, in which the brain is connected to peripheral nerves and muscles along a route ventral to (beneath) the enteron, in Chordata this central route runs dorsal to (above) the enteron, within a *notochord*. In higher chordates, the notochord became bony and rigid, and this structural backbone, in combination with the six-celled brain, gave chordates a particularly fast form of locomotion.

A simple, bilateral vertebrate like a fish moves forward by successively contracting first the muscles on the left side of its backbone, then those on the right, then the left, and so on. But to get this rhythm, the minimal anatomy of the chordate brain needs *two* two-celled brains—one for each side of its body—as well as two *inhibitory* cells like $R_i$ and $L_i$ in figure 2.9a. This brain, however, is still "convulsive" and slow like a jellyfish. Neither $L_e$ nor $L_i$ will stop signaling until it has exhausted itself. Only then will it become possible for $R_e$ to activate and drive the fish's *contralateral* ("other-side") stroke.

The anatomy of figure 2.9a will reappear time and again in subsequent chapters, where its resonant architecture will be used to drive learning. But to survive in the primordial soup, the first vertebrates had to be fast and efficient, and for this they had to give the anatomy of figure 2.9a a further twist. In figure 2.9b, the motor drive signals from $L_e$ and $R_e$ are twisted contralaterally.[4] In chapter 4, we will use this twist to associate $L_i$ and $R_i$ with the modern vertebrate cerebellum. For our ancient ancestors, however, it was enough that $L_e$

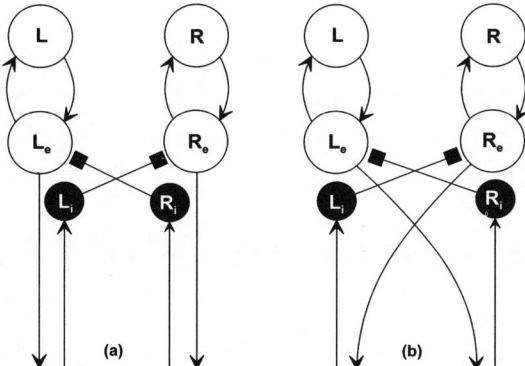

**Figure 2.9.** A six-celled brain: bilateral symmetry with inhibition and proprioceptive feedback. (a) Sensory circuit. (b) Motor circuit.

drove the right side of the body and that the motor command system formed a loop through $R_i$. When proprioceptive signals from the fish's right side told $R_i$ that the fin stroke was complete, $R_i$ inhibited $L_e$, conserving energy and allowing $R_e$ to activate its stroke on the left side of the fish's body. Thus, the vertebrate swam through the water, rhythmically swinging its tail from side to side, the quickest predator in the Ordovician sea.[5]

• T H R E E •

# The Communicating Cell

In the last chapter we reviewed some of the problems one-celled life solved in its struggle to survive. In order to get ahead in life, we saw that the primeval organism needed to grow and to move. This led to multicelled creatures with the new problem of intercellular communication. To get ahead, the multicelled organism's left cell had to know what the right cell was doing. Thus, the evolutionary differentiation of cells leads us to the origin of mind: knowing what one is doing.

For muscle cells to work together, they must be coordinated. This implies that some cell, or group of cells, must take charge and communicate an order to muscle cells, many of which are relatively distant. By what structure and process could one cell communicate with another cell over a distance? In the modern neuron, it is the *axon* that makes this possible. The axon is a long, thin, tubular extension of the cell's membrane that carries electrical intercellular communications. There is no clear fossil record of how the axon evolved, but as we saw in chapter 2, an obvious candidate prototype was the Mastigophora's flagellum: Nature having once invented a distal extension of the cell, nature did not have to reinvent it. Nature only had to remember it somewhere in DNA and then adapt it to create a new cell type, the neuron.

Since the functions of brain cells were a mystery to early anatomists, different neurons were first named by the shapes of their cell bodies, and in the first half of the twentieth century, a menagerie of "pyramidal" and "spherical," "stellate" and "bipolar," and "spiny" and "smooth" neurons was collected under the microscope. But whatever their body shape, all neurons have long axons.[1]

The large pyramidal cells of cortex (from the Latin for "rind") stain particularly well and became the early objects of microscopic study. Cerebral cortex is a sheetlike fabric of neurons about 4–5 mm thick. In microscopic cross sections of this fabric like figure 3.1, early researchers could see the apical

**Figure 3.1.** The laminae of cerebral cortex. (Lorento de Nó 1943. Reprinted by permission of Oxford University Press.)

dendrites of pyramidal cells rising high above their cell bodies, finally spreading out, in treelike "arborizations," while lesser, basal dendrites spread out from the bottom. Below the cell body, axons could be seen descending below the cortical sheet (*a, b, c,* and *d* in figure 3.1). But where do these axons go? They quickly outrun the microscope's field of view. If one could follow figure 3.1 several frames to the right or the left, it would become clear that many of the axons arise again into the cortical sheet, where they connect with the dendrites of other cortical neurons. But it remains almost impossible to know exactly which neurons connect with which other neurons.

This problem becomes even worse if one looks very closely at some of the pyramidal cells in figure 3.1. There it can be seen that axon collaterals branch off and radiate from the main axon. So to learn the connections of any neuron, we must trace not just one axon but thousands of axon collaterals. That there are thousands of axon collaterals for every main axon is graphically demonstrated in figure 3.2, a drawing from an electron micrograph of an average neuron's cell body. Each bump on the cell body is the synaptic connection of some axon collateral. If the average neuron receives thousands of connections, as in figure 3.2, then it follows that an average neuron must also send out thousands of axon collaterals, each originating from a single inconspicuous output fiber.

To put the problem in further perspective, imagine that a largish pyramidal cell 150 μm in diameter was actually a large tree with a trunk 1 m in diameter. Then the tree's apical dendrite would rise 50 m above the ground, and its basal dendrite "root system" would be 30 m in diameter. All of this relates nicely

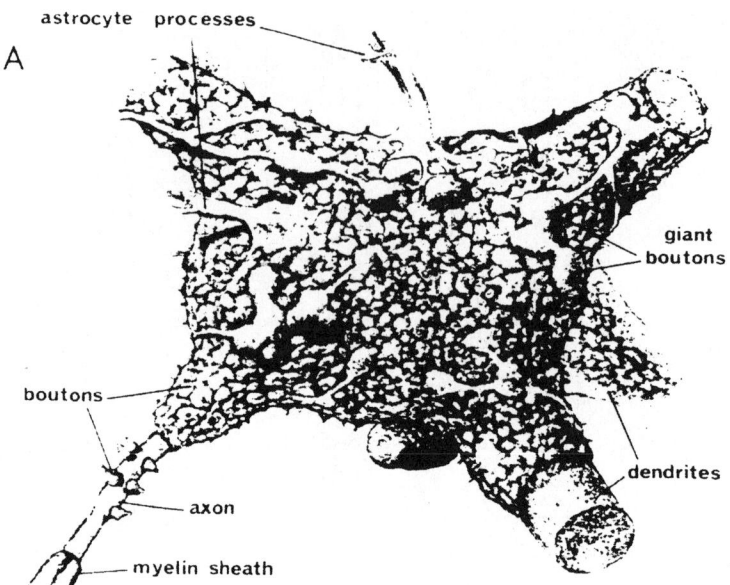

**Figure 3.2.** Competition for synaptic sites is intense. (Poritsky 1969. Reprinted by permission of John Wiley and Sons.)

to the proportions of a large tree. But the main axon collaterals would run a distance of 1 km! Making the problem worse still, many main axon collaterals descend and run this distance beneath the cortex in a great labyrinth of "white matter," a tangled mass of trillions of other axons, a cortical "underground" (a, b, c, and d in figure 3.1). Now imagine the task of excavating and tracing a single axon. Sometimes, the axons form bundles, called *fascicles*, which, like a rope, can more easily be traced from brain region to brain region. In this way we know, for example, that a bundle of axons (the *arcuate fasciculus*) connects Broca's area and Wernicke's area. Only within the last decade has science finally begun to accurately trace detailed pathways from neuron to neuron using radioactive and viral tracing techniques, and even still the problem is daunting.

It is no wonder that throughout the twentieth century, researchers by and large ignored the axon collaterals and instead focused their ever-more-powerful microscopes on ever-smaller neural structures in an ever-narrowing field of view. We shall do the same in this chapter. We shall focus on details of neural structure. But in the rest of this book, our main task will involve understanding the broad view, the connectivity patterns of axon collaterals.

## The Neuron Membrane

Toward the end of chapter 2, we saw that the early neuron had to develop a cable to communicate its messages to a muscle cell, and we identified the axon as that cable. But how exactly does a message travel down the axonal cable? Since axons are hollow, one might imagine chemicals diffusing or even (à la Descartes) being pumped through the axon's interior. A moment's reflection on the tree analogy and the 1 km axon should convince us that this would be a hopelessly slow mechanism. And blessed with the hindsight of twentieth-century science (and Galvani's eighteenth-century observations), we know that the nervous signal is electrical. Where could the first neuron have gotten the idea of electrical communication?

Thinking back to chapter 2, recall the membrane feeding frenzy. There we postulated that the "mouths" of the protozoan cell membrane could communicate via the ionic charges of sodium ($NA^+$) and chloride ($Cl^-$) in the primordial soup. The evolution of nervous signaling through recruitment of such an ionic feeding signal is quite speculative, but there is no longer anything speculative about the role of $Na^+$ and $Cl^-$ in the propagation of nervous signals. In 1963 Hodgkin and Huxley received the Nobel Prize for defining the role of $Na^+$ and $Cl^-$ in nerve signal propagation along the giant axon of the squid.

Not all nervous systems evolved exactly like the human one. Mollusks developed along a rather different line, and squid evolved axons up to 500 mm in diameter, some 100 times larger than a comparable vertebrate axon. A series of twentieth-century studies based on these giant squid axons established many essential facts about how nerve cells transmit signals along the axon. Into such large axons, Hodgkin and Huxley were able to insert microelectrodes and micropipettes to measure differences in charge and chemical concentrations inside and outside the axon. Living cells tend to be negatively charged, and in most neurons this charge is usually expressed as an internal, negative, "resting-level" charge on the order of −70 millivolts, relative to the surrounding plasma. This can be almost entirely attributed to a lower internal concentration of positively charged sodium ions.

Figure 3.3 schematically depicts a segment of an axon membrane and the process Hodgkin and Huxley discovered. In the figure, when a sodium gate is opened, positively charged $Na^+$ ions are drawn through the membrane into the cell interior. This local voltage drop (a "shock," or *depolarization*) causes adjacent sodium gates to open, and the sodium influx moves along the axon membrane. This simple $Na^+$ chain reaction is the basis of the "nervous" signal, but it has its limits. If sodium is allowed to flow into the axon unchecked, the resting potential of the membrane will soon reach 0 mV. Once this happens (and actually well before this happens), another signal cannot be generated until the excess sodium is somehow pumped out of the axon and the original −70 mV resting-level charge is restored. In fact, cells do have membrane "waste" pumps which remove excess sodium and other waste from the cell interior, but these pumps are metabolic and operate much more slowly than the electrical forces of ions and the nerve signal.

The successful neuron could not wait around for these bilge pumps. The nerve cell membrane therefore evolved a separate set of potassium gates to control the influx of sodium. As figure 3.3 illustrates, when an initial $Na^+$ in-

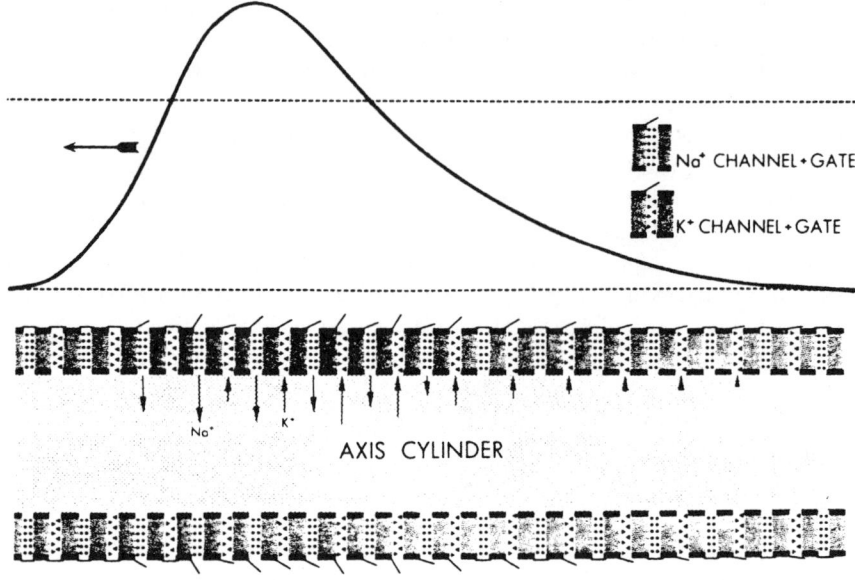

**Figure 3.3.** Sodium and potassium gates. (Eccles 1977. Reprinted by permission of McGraw-Hill Book Company.)

flux locally depolarizes the axon membrane, Na⁺ rushes in. But in the neuron, the same ionic forces cause potassium gates to open. Positively charged potassium ions (K⁺) leave the cell, repolarizing the membrane and closing the open Na⁺ gates. Potassium being heavier than sodium, we may imagine the potassium gates and ions as being relatively ponderous and slow. Thus, the K⁺ current is exquisitely timed to stop the chain reaction just after the sodium charge has begun to propagate down the axon but before an influx of Na⁺ floods and discharges the membrane unnecessarily. The time course of these events is plotted in figure 3.4.

On an oscilloscope, as in the figure, this brief membrane depolarization shows up as $V$, a brief voltage *spike*, or *action potential.* After the spike, metabolic pumps must still restore K⁺ and Na⁺ to their original levels, but this is now a minimal task, since only enough Na⁺ and K⁺ was transported across the membrane to generate a single spike. In the meantime, the −70 mV resting-level charge (−64 mV in the particular cell measured in figure 3.4) can be speedily restored, and a new Na⁺ pulse can be generated and propagated. Most central nervous system (CNS) neurons have a "refractory time" on the order of 2.5 ms. During this time, the neuron cannot generate another spike. This means that most neurons can generate spikes as frequently as 400 times per second, but there is substantial variation. Renshaw interneurons,[2] for example, have been found to fire up to 1,600 times per second.

By implanting microelectrodes at two points along the axon in figure 3.3, it is possible to electronically measure the speed at which a spike propagates.

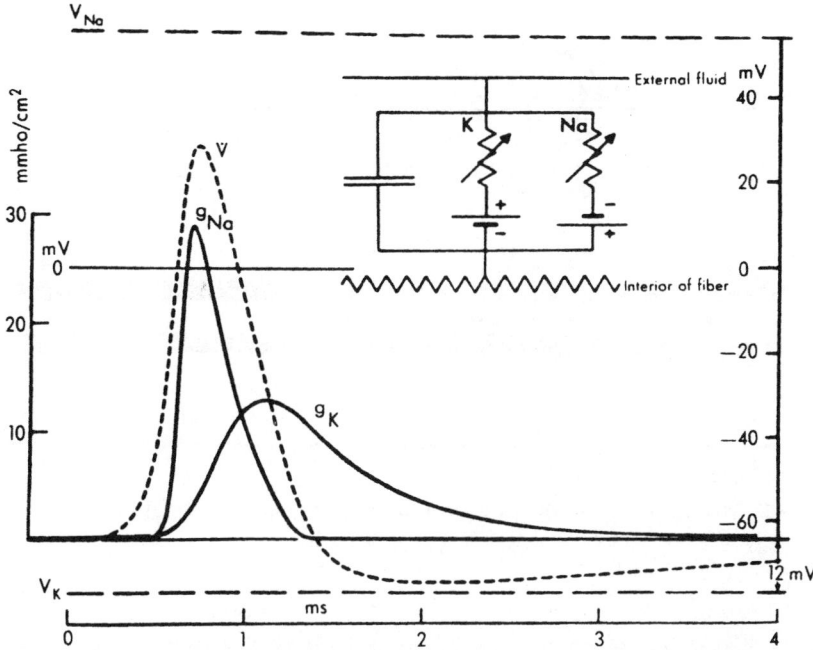

**Figure 3.4.** Sodium and potassium currents create and limit the duration of signal "spikes." (Eccles 1977. Reprinted by permission of McGraw-Hill Book Company.)

As a rule of thumb, nerve signals travel at a rate of 1 m/s, but a typical speed for a 30 μm invertebrate axon might be 5 m/s. The speed is directly proportional to the diameter of the axon. This explains why the squid, as it evolved to be larger, evolved a giant axon to signal faster at larger and larger distances. With a 500 μm axon, the squid's nervous signal travels at 20 m/s. A little math, however, will show that the squid's solution is something of an evolutionary dead end. There are limits to growth. The speed of transmission is proportional to the *square root* of the axon diameter, but the axonal volume which the cell metabolism must support grows with the *square* of the diameter. The squid was caught in a game of diminishing returns. Vertebrates found a better solution.

## Myelin

Vertebrates evolved a type of cell that wraps itself around an axon. In the brain, these cells are called oligodendrocytes, whereas elsewhere, they are called Schwann cells, but in both cases the resultant axon wrapping is called a myelin sheath or, simply, *myelin*. Between the sheaths of successive oligodendrocytes, the axon is exposed at "nodes of Ranvier" (figure 3.5). When a sodium influx depolarizes the extracellular fluid at such an exposed node, adjacent sodium gates cannot be opened, because they are under the myelin sheath. Instead,

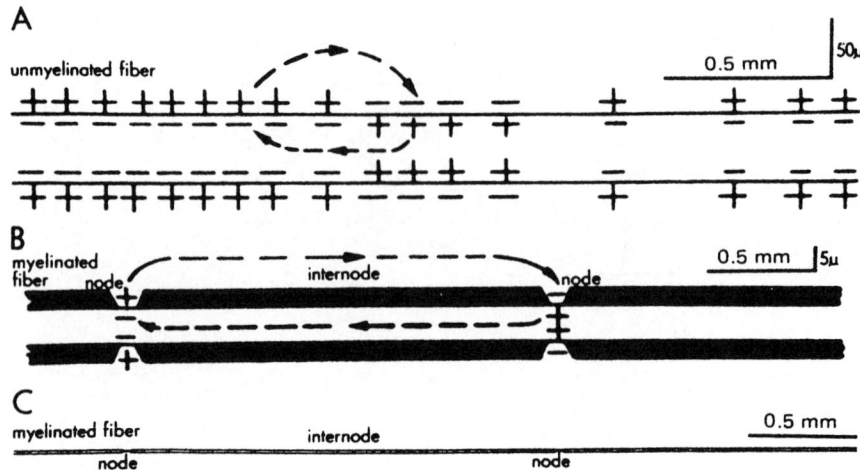

**Figure 3.5.** Myelin. (Eccles 1977. Reprinted by permission of McGraw-Hill Book Company.)

the ionic influence is exerted on the next node of Ranvier, opening sodium gates there. As a result, nervous impulses "jump" from node to node at electrical speeds which are not limited by the mechanical opening and closing of local membrane gates. This process is therefore sometimes called saltatory conduction, from the Latin *saltare* < *salire*, "to leap." So while a 500 μm giant squid axon labors to achieve speeds of 20 m/s, a large, 5 μm, myelinated vertebrate axon can achieve speeds of 120 m/s. As a bonus, the myelinated neuron also needs far fewer ion pumps to restore ionic balance after each pulse and so uses less metabolic energy. The myelin sheath also provides structural support for the axon, allowing it to be thin and more energy-efficient without loss of strength.

Figure 3.6 illustrates how successive myelin cells envelop an axon. Myelin is white in appearance, so it is the myelin sheaths which give the mass of axons beneath the cortex the name "white matter." Inflammation and/or degeneration of myelin sheaths cause major neural pathways to slow down to a relative halt. This is the debilitating disease known as multiple sclerosis.

## Thresholds, or Why Neurons Are (Roughly) Spherical

Although I have explained how the nervous signal came to be, and how it came to be fast, I have not explained how a nerve signal, being just barely sustained along the surface of a 3 μm axon, could depolarize a 150 μm cell body.

It can't. So where the axon terminates on another cell, the presynaptic axon terminal swells, forming a larger synaptic *terminal*, or "knob." The postsynaptic cell also swells at the contact site, forming a *spine* (figure 3.7). Now the

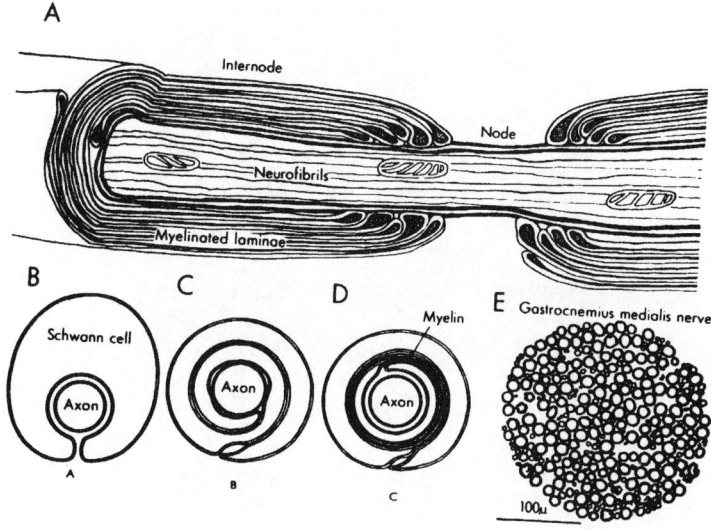

**Figure 3.6.** Oligodendrocytes and Schwann cells wrap axons in myelin. (Joseph 1993. Reprinted by permission of Plenum Press.)

impedance of the axon more nearly matches the impedance of the postsynaptic membrane. When the charge does successfully cross the synapse onto the postsynaptic membrane, the spine funnels the charge onto the postsynaptic cell body.

Yet it is not funneled directly onto the cell body. As figures 3.7 and 3.2 illustrate, the spine is but the smallest branch in an arborization of dendrites, which resembles nothing so much as our fractal fern in figure 2.4. Thus, a single spike never depolarizes a postsynaptic cell body by itself, but if one small dendritic branch is depolarized just as it meets another small, depolarized branch, then the combined charge of the two together can depolarize a dendritic limb. And if that depolarized limb should meet another depolarized limb, then that limb will be depolarized, and so on, until the "trunk" of the dendritic tree builds a charge sufficient to begin depolarizing the cell at the "north pole" of the cell's spherical body.

In order for the depolarizing current to reach the axon, a larger and larger band of depolarization must spread out from the north pole—until the depolarization finally reaches the cell's equator. From there to the south pole, the depolarization spreads to an ever-decreasing area of membrane. Thus, the equator defines a threshold, a degree of membrane depolarization which must be exceeded if depolarization is ever to reach and propagate along the axon. Once exceeded, however, a coherent charge, limited by the dynamics illustrated in figure 3.4, is delivered to the axon.

A generation of researchers was led astray along with Von Neumann 1958 in concluding that since the spike was a discrete, thresholded, "all-or-nothing"

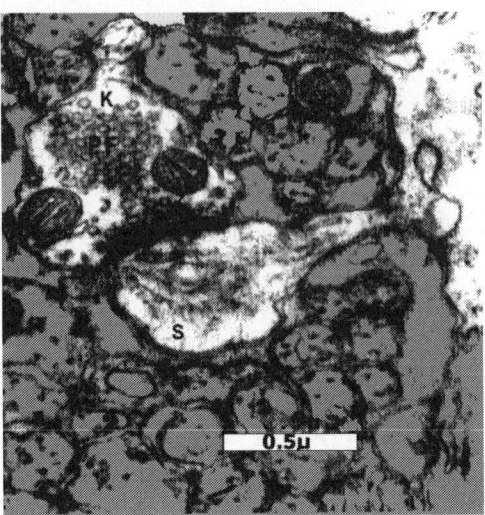

**Figure 3.7.** Photomicrograph of a synapse. A highlighted dendritic spine protrudes downward from right-center, with a highlighted axon terminal knob synapsing from above. (Llinás and Hillman 1969. Reprinted by permission of the American Medical Association.)

event, the nervous system itself was a discrete, binary system like a computer. If we compare the surface area of the axon to the surface area of the cell body, we realize that the thin axon cannot carry away the entire charge of the cell's southern hemisphere in one spike. Instead, a *volley* of spikes is normally released, one after another. The frequency of spikes in this volley varies in proportion to the charge on the cell body, up to the limit imposed by the refractory period of the cell membrane, and this variable spiking frequency carries much more information than a simple 1 or 0.

Traveling away from the cell body, the spike eventually brings us back to the problem Golgi tried to explain to Ramón y Cajal (chapter 1). If the cell membranes are not connected, how can the charge be transferred across this physical gap, called a *synapse*, onto the postsynaptic membrane?

## The Synapse

At its end, each axon collateral abuts (but is not physiologically attached to) another cell's membrane. This junction is called a synapse. Even after solving the impedance mismatch problem with a system of terminal knobs, spines, and dendrites, there remains the biological hurdle of communicating the nervous signal from the presynaptic cell to the postsynaptic cell, across a synaptic gap. The quickest way to pass a nervous impulse would be to simply pass the ionic chain reaction directly onto the postsynaptic cell membrane. In fact "gap junctions" are commonly found in submammalian species like electric fishes. A few are even found in the human nervous system, but they are very rare. After all, if Na⁺ flows into the presynaptic terminal knob when it is depolarized, then there will be that much less Na⁺ left in the synapse to depolarize the postsynaptic membrane. Instead, at virtually all synapses, specialized chemical effluents from the presynaptic axon terminal—*neurotransmitters*—open specialized pores in the postsynaptic cell membrane, and the opening of these gates reinitiates depolarization. But this is a slow process. As Sherrington demonstrated in 1906 (incidentally disproving Golgi's continuous-network hypothesis and winning Ramón y Cajal's side of the argument), it decreases the speed of the neural signal by a factor of ten. If the race of life goes to the quick, how did the chemical synapse survive?

As it happens, in the mid-1930s electrical engineers began to model neural circuits rather as if they were Golgian radio circuits. In a famous paper McCulloch and Pitts (1943) offered networks like those in figure 3.8. Like the last of these, which features a reverberatory loop, such networks had many interesting properties, but they had one glaring pathology: when the current is switched off or interrupted in such a network, all memories are lost.

Donald Hebb noted that "such a trace [a McCulloch-Pitts reverberatory loop] would be unstable" (Hebb 1949, 61). If the current were turned off, the thought would be lost. Such Golgian "gap junction" synapses have no mechanism for long-term memory. Where then might long-term memory be found in a real brain? Hebb went on to speculate that long-term memory (LTM) must therefore reside at the (chemical) synapses, so that "when the axon of cell A is near enough to excite a cell B and repeatedly or persistently takes part in firing it, some growth process or metabolic change takes place in one or both cells such that A's efficiency, as one of the cells firing B, is increased. . . . A re-

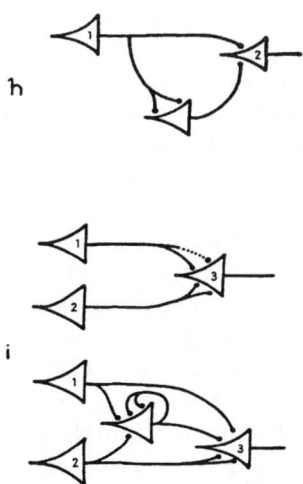

**Figure 3.8.** A McCulloch-Pitts neural network. (McCulloch and Pitts 1943. Reprinted by permission of Elsevier Science Ltd.)

verberatory trace might cooperate with the structural change and carry the memory until the growth change is made" (1949, 61–62).

Thus, Hebb described synaptic learning as a physiological associative process. Since associative learning had been extensively developed as a psychological concept (Ebbinghaus [1913] 1964), Hebb's criticism of the McCulloch-Pitts model found a wide and receptive audience among behaviorists. The Hebbian theory joined both learning and long-term memory at the synapse, and soon microscopic evidence was found which tended to support his conjecture. It was found that disused synapses tended to atrophy, so the inverse seemed plausible: used synapses would hypertrophy—they would grow. Enlarged axon terminals would have more neurotransmitter and would therefore engender larger depolarizations in their postsynaptic spines, which, being larger also, would pass larger nervous impulses onward. This makes for a very clear picture of LTM, and indeed, we will use hypertrophied presynaptic knobs to represent LTM in the diagrams which follow throughout this book. However, the actual synaptic mechanisms which underlie long-term memory are more complicated, and more wonderful. For a better understanding of the modern theory of synaptic learning and memory, we must look more closely at neurotransmitters.

## Neurotransmitters and Membrane Receptors

I have already noted that axon terminals form "buds" or "knobs" which increase the contact area between the otherwise very fine axon and the much larger target cell body. Each of these knobs is like a small branch office of the main neuron cell body. Within each knob is an almost complete set of mitochondria to power the branch office and organelles to manufacture locally essential chemicals. All that is really lacking to make the knob a self-sufficient neuron is self-replicating DNA and the cell nucleus—a minor difference since mature neurons seem to reproduce rarely. Inside the synaptic knob, small vesicles of neurotransmitter (the small bubbles in figure 3.7) accumulate along the membrane adjacent to the synaptic cleft, ready for release. When the membrane depolarizes, the vesicles empty neurotransmitter into the synaptic cleft.

The number of neurotransmitters identified has grown substantially since Loewi first identified acetylcholine as a neurotransmitter in 1921. There are now about a dozen known primary neurotransmitters and several dozen more *secondary messengers*. But what has proved more complex still is the variety of neurotransmitter *receptors*. When an axon terminal releases neurotransmitter, the neurotransmitter itself does not penetrate the postsynaptic membrane. Rather, it attaches to receptor molecules in the postsynaptic membrane. These receptors in turn open channels for ions to flow through the postsynaptic membrane. There are several and often many different receptors for each neurotransmitter. Apparently, a mutation that changed the form of a neurotransmitter would have broadly systemic and probably catastrophic consequences, but mutations in the structure of receptors are more local, more modest, and have enabled more nuanced adaptation.

Over a dozen different receptors have been identified for the neuromuscular transmitter *acetylcholine*, and over twenty different receptors in seven distinct families (5–HT1–7) have been identified for the neurotransmitter *serotonin* (also known as 5–HT, 5–hydroxytryptamine). Some receptors admit anions ($Na^+$ and $Ca^{2+}$), depolarizing and exciting the postsynaptic cell. Other receptors admit cations (principally $Cl^-$), hyperpolarizing and inhibiting the postsynaptic cell. Thus, one should perhaps no longer speak of excitatory and inhibitory neurotransmitters, since many can be either, depending upon which type of receptor the neurotransmitter binds to. This is especially true of neurotransmitters that bind to G-protein-coupled receptors: notably *dopamine, serotonin*, and the adrenergic transmitter *noradrenaline*.[3] In such synapses, regulatory G-proteins and catalysts like adenylate cyclase (cAMP, adenosine 3',5'-cyclic monophosphate) act as *second messengers*, further modulating membrane polarization. As we shall see, these have various and complex effects upon the CNS, and they have received considerable recent attention. Drugs like Ritalin™ widely prescribed for attentional deficit disorder) and Prozac™ widely prescribed for depression and anxiety) affect the adrenergic and serotonergic systems, respectively. Dopamine deficiency has been isolated as a cause of Parkinson's disease, and Gilman and Rodbell were awarded the 1994 Nobel Prize for initially elucidating the function of G-proteins in neurobiochemical signaling.

The other major group of neurotransmitters are those that effect fast signaling directly through ligand-gated ion channels. Fortunately, these neurotransmitters tend to be more uniformly excitatory or inhibitory, and for our minimal anatomies, we need focus on only two: *glutamate* (and its chemical cousin *aspartate*) and *gamma-aminobutyric acid* (GABA).

Glutamate is an excitatory neurotransmitter. It is produced by major brain cells like the pyramidal cells of neocortex, and it induces depolarization in its target membrane, as described above. In the main excitatory case, glutamate is released from the presynaptic axon terminal and attaches to a receptor gate on the postsynaptic membrane.

In 1973, Bliss and Lømo, using microelectrodes, studied the "evoked postsynaptic potential" response of hippocampal pyramidal synapses to repeated stimulation. For the first few intermittent stimulations, a modest response was obtained. But after a few of these stimulations, the synapse began to generate a bigger and bigger response to the same stimulation. The synapse seemed to learn. Moreover, this learning effect persisted. Called long-term potentiation (LTP), this effect is now the leading physiological explanation of long-term memory. In figure 3.9, the effects of LTP are graphed. First, several strong, "tetanizing" stimuli are applied to a postsynaptic membrane until $t = 0$. For a long time thereafter, milder stimuli continue to elicit an elevated response.

As it turns out, glutamate attaches to two distinct types of receptors: *N-methyl D-aspartate* (NMDA) sites and non-NMDA sites.[4] At first, in "normal" transmission of the nerve signal, glutamate opens the non-NMDA gates, allowing $Na^+$ to enter the postsynaptic cell and reinitiating our familiar $Na^+/K^+$ chain reac-

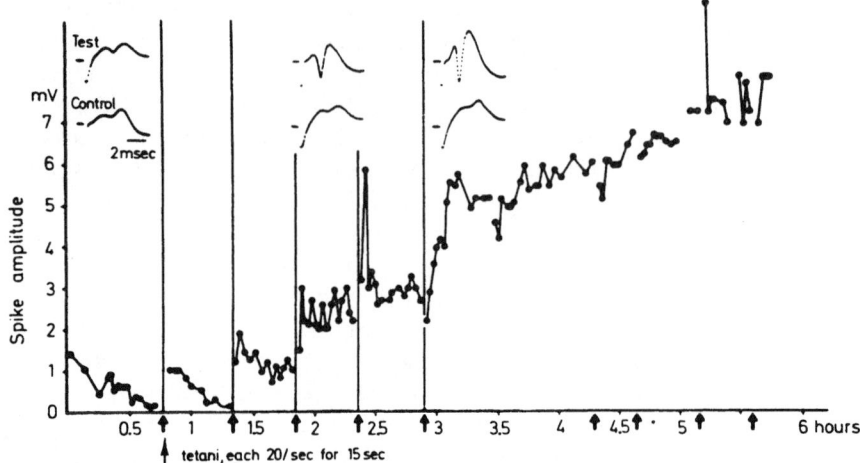

**Figure 3.9.** Long-term potentiation. (Bliss and Lømo 1993. Reprinted by permission of the Physiological Society.)

tion. It turns out that glutamate also attaches to NMDA gates. These gates are initially blocked by $Mg^{2+}$ ions, but as the membrane depolarizes through continued opening of non-NMDA receptor gates, the $Mg^{2+}$ ions are dislodged. Once opened, the larger NMDA receptors admit doubly charged calcium ($Ca^{2+}$) into the cell interior. There the calcium also interacts with calmodulin and CAM kinase II, which phosphorylates the non-NMDA glutamate receptors, causing them to be more readily activated in the future. This process is believed to be the basis of the physiological process of long-term potentiation and of the psychological phenomenon of LTM (Grossberg 1968; Lynch 1986; McGlade-McCulloh et al. 1993).

GABA, on the other hand, is inhibitory. It is released by cortical interneurons like basket cells and chandelier cells (figure 3.10). Whereas glutamate opens gates for positive ions, GABA opens channels for the entry of negative ions, principally Cl−, into the cell. This influx of chloride ions *hyperpolarizes* the postsynaptic membrane, from −70 mV to perhaps −80 mV, thereby making it more difficult for excitatory neurotransmitters like glutamate to depolarize the membrane and initiate spiking. This hyperpolarization of cells does not have to match impedance in order to propagate a spike, so inhibitory cell synapses do not exhibit spines, and they are more commonly found on cell bodies than in dendritic arbors.

For example, GABAergic chandelier cells synapse preferentially on the initial axon segments of neocortical pyramidal cells. When stained as in figure 3.10, the terminal knobs of their axon collaterals look like so many candlesticks, hence their name. The axon terminals of basket cells, on the other hand, form a "basket" of synapses on the bodies of neighboring excitatory cells (see figure 4.8).

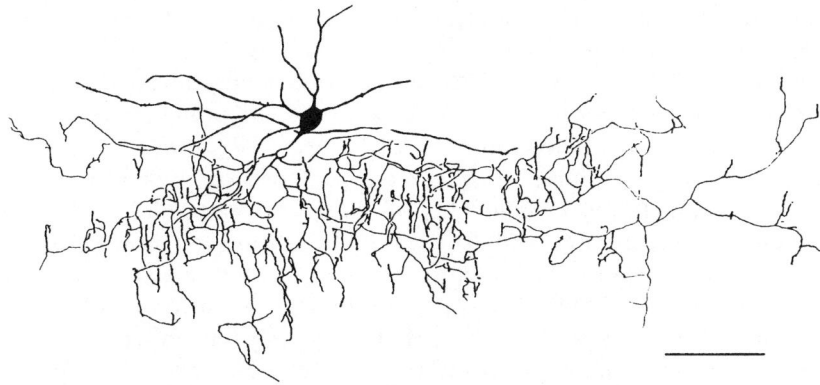

**Figure 3.10.** A chandelier cell. (A camera lucida drawing from Peters and Jones 1984, 365. The bar equals 25μm. Reprinted by permission of Plenum Publishing Corp.)

By some reports, inhibitory neurons compose less than one-fifth of the total number of CNS neurons, but Crick and Asanuma (1986) observe that even if inhibitory cells are outnumbered, they may exercise a disproportionate "veto" power over excitatory signals by exerting their inhibitory influence on prime real estate like cell bodies and initial axon segments.

As I have noted, numerous other neurotransmitters are present in the CNS in lesser quantities. The most important among these are *acetylcholine, noradrenaline*, and the G-protein neurotransmitters *serotonin* and *dopamine*.

Acetylcholine is the neurotransmitter with which motor neurons cause muscles to contract and is apparently the principal neurotransmitter of the parasympathetic (smooth-muscle) nervous system. Acetylcholinergic fibers are also found widely distributed in neocortex and the midbrain. These seem to mostly arise from the reticular formation of the brain stem, part of the system for perceiving pain. Because it is present and easily studied at the neuromuscular junction, acetylcholine was the first neurotransmitter to be identified. For his discovery of the neurotransmitter role of acetylcholine, Loewi received the 1936 Nobel Prize.

Serotonin has been traditionally classed as an inhibitory neurotransmitter, although it has recently been found to also exert an excitatory effect upon cerebral cells (Aghajanian and Marek 1997). Since serotonin levels are selectively elevated by antidepressives like fluoxetine (Prozac), serotonin has become famous for its role in controlling various mood disorders.

Noradrenaline is another widespread neurotransmitter. It is identical to what elsewhere in the body is called the "hormone" *adrenaline*, but adrenaline does not cross the blood-brain barrier. The brain must manufacture its own supply of adrenaline, which it does from dopamine. In the brain, this adrenaline is called *noradrenaline* or *norepinephrine*. In the hippocampus of the brain, and presumably also in neocortex, which develops from the hippocampus, noradrenaline has a complex effect. It inhibits neuronal response to brief stimuli but increases neuronal response to prolonged stimuli.

Figure 3.11 illustrates this differential effect of noradrenaline on inputs to a hippocampal pyramidal neuron. In the first case shown in figure 3.11A, a small, ramped input (below) elicits a spike response (above). By contrast, a longer, more intense "pulse" input elicits a burst of seven spikes in response to the pulse onset. However, under the influence of noradrenaline, the ramped input elicits no response, while the pulse elicits fourteen spikes throughout the duration of the pulse. That is, noradrenaline inhibits small inputs but makes the neuron hyperexcitable and responsive to the more intense pulse stimulus. Madison and Nicoll (1986) observe that this effectively enhances the signal-to-noise ratio of the neuron's response. I will elaborate on such effects later.

Dopamine is a CNS neurotransmitter which is chemically transformed into noradrenaline (both are *catecholaminergic*). Dopamine deficiency has been found to be a symptom of Parkinson's disease. Treatment with L-dopa, a dopamine precursor, with subsequent increases in dopamine levels in the subcortical

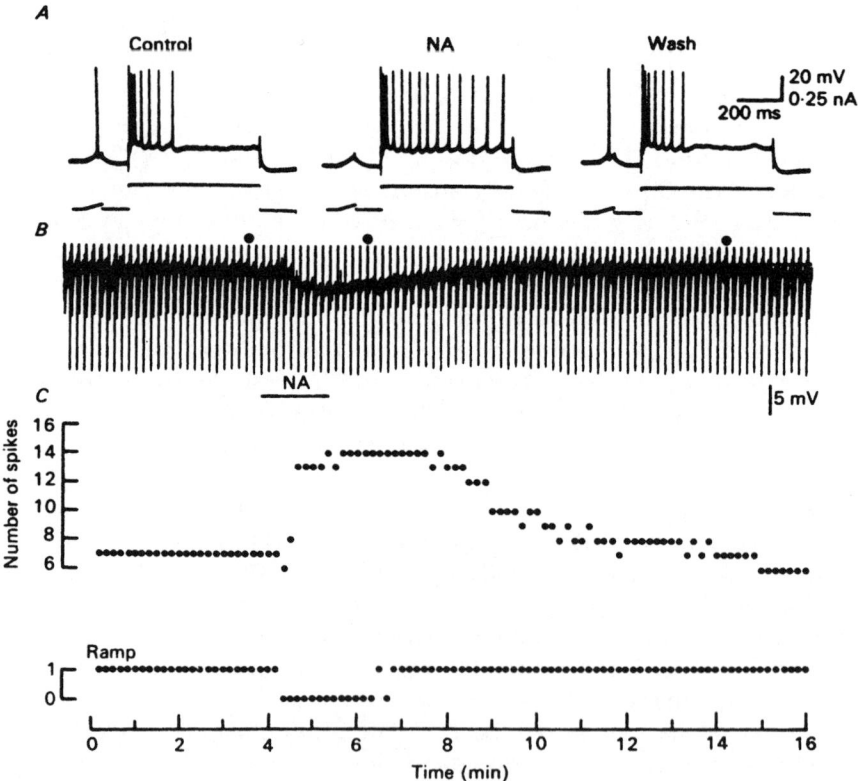

**Figure 3.11.** Effect of noradrenaline on evoked postsynaptic potentials. (Madison and Nicoll 1986. Reprinted by permission of the Physiological Society.)

basal ganglia, described in the following chapter, has ameliorated the symptoms of Parkinson's disease for many patients.

## Reverse Messengers

At the same time that neurotransmitter released from the presynaptic knob acts upon the postsynaptic cell, *retrograde* neurotransmitters such as *nerve growth factor* (NGF; Thoenen 1995) and *nitric oxide* (Kandel and Hawkins 1992; Snyder and Bredt 1992) are released from the postsynaptic cell and act upon the presynaptic cell, opening $Ca^{2+}$ gates and otherwise facilitating the metabolism of the presynaptic knob. A kind of intercellular free-trade agreement is set up, and the economies of both the presynaptic cell and the postsynaptic cell begin to grow (Skrede and Malthe-Sorenssen 1981; Errington and Bliss 1982).[5]

Thus, the cell membrane, life's first defense in the primordial soup, evolved to allow cooperative commerce among friendly neighbors. There are many examples of symbiosis in nature, but the cooperative exchange of ions between two protoneurons was certainly one of its first and most significant occurrences. Once two neurons could communicate between themselves, it was relatively easy for 200 or 2,000 to organize themselves in patterns of self-similarity. It took something like a billion years for life to progress to the two-celled brain. In half that time, the two-celled brain evolved a million species, including *Homo loquens* with his ten-trillion-celled brain, to which we now turn in chapter 4.

• F O U R •

# The Society of Brain

In time, the primitive brains that had developed by the end of chapter 2 surrounded themselves with yet another self-similar membrane. Called the *notochord*, this cartilaginous tube houses the primitive central nervous system of the phylum Chordata. In ontogeny as well as phylogeny, this notochord further develops into a hard, tubular backbone, so the members of Chordata have come to be popularly known as "vertebrates." The long, hollow notochord/backbone with its bulbous skull at one end is self-similar to the long, hollow axon and its cell body, and it serves the self-same purpose of communication in these larger, more complex organisms. In time, the four-celled model brain of chapter 2 evolved into the trillion-celled human brain.

Elephants and whales have larger brains than humans, and they are probably more intelligent, too. At least they have never engaged in such colossal stupidity as World War I. But if we must find a way to assert humankind's supposedly superior intellect, we can observe that elephants and whales seem to devote a good portion of their prodigious brains simply to moving their prodigious bulks, and we can note the oft-repeated fact that per kilogram of body, *Homo loquens* has the largest brain in the world. But even this measure does little to support our vanity. Among mammals, birds, and reptiles, there is only a 5% variation in the ratio of brain size to body size (Martin 1982). The proper question therefore seems to be, not how *big* an animal's brain is or how "intelligent" the animal is, but what it does with the brain it has.

## Subcerebral Brain Structures

If we peer beneath the surrounding cerebrum, we see the older structures which the cerebrum has overgrown (figure 4.1). These subcerebral (or "subcortical")[1] structures are usually not considered specialized for cognition or language, but

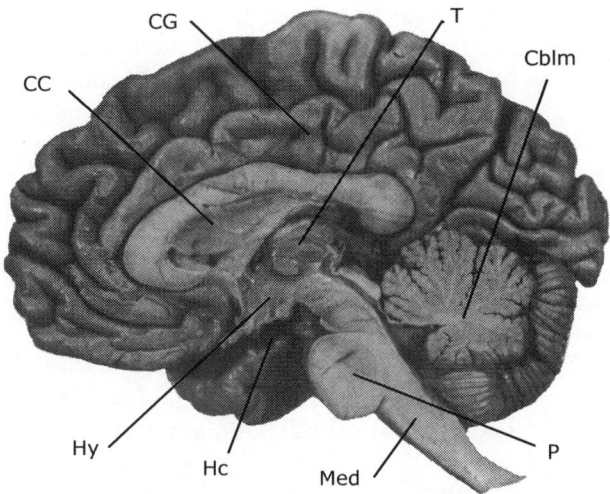

**Figure 4.1.** Medial section of the human brain. (CC) Corpus callossum; (CG) cingulate gyrus; (T) Thalamus; (Cblm) cerebellum; (P) pons; (Med) medulla; (Hc) hippocampus; (Hy) hypothalamus. (After Montemurro and Bruni 1988. Reprinted by permission of Oxford University Press.)

they are nonetheless indispensable to higher cortical functioning, so at least a quick survey is in order.

At the base of the subcerebral brain structures is the *brain stem*. Like the primitive notochord, it can be viewed as the source of all further brain development. The lower part of the brain stem, the *medulla* (Med in figure 4.1) conducts *afferent* (ascending) signals from the periphery to the brain and *efferent* (descending) signals from the brain to the periphery. These peripheral connections can be either relatively local (e.g., to the eyes, ears, and mouth) or quite distant (e.g., to the limbs via the neuromotor highway of axons called the *pyramidal tract*).[2]

Several diffuse catecholaminergic neurotransmitter systems arise from the medulla that have important connections to the limbic system and the hypothalamus. These include important noradrenaline (norepinephrine) systems arising from the *locus coeruleus*, and dopamine systems arising from the *substantia nigra*. A serotonergic system arises from the *raphe nuclei* of the medulla, also with diffuse connections in the limbic system and cerebrum. These systems modulate behavior, but in globally autonomic and emotional ways rather than through cognition.

Rostral to (i.e., above, toward the head of) the medulla is the *pons* (P in figure 4.1). The pons relays ascending and descending motor signals and is the primary relay site for signals to and from the cerebellum. Above the pons is the *midbrain*, a short section of the brain stem containing the *superior* and *inferior colliculi*. The superior colliculus is a major relay point in the visual system, while the inferior colliculus is a major relay point along auditory pathways.

Rostral to the midbrain is the *diencephalon*. We divide the diencephalon into three parts: the *thalamus*, the *basal ganglia*, and the *limbic system*. The *thalamus* (T in figure 4.1) receives afferent (ascending, incoming) sensory signals and relays them to the cerebrum. Within the thalamus, the *lateral geniculate nucleus* (LGN) is an important relay point for visual signals, while the *medial geniculate nucleus* (MGN) is the thalamic relay point along auditory pathways. Most afferent sensory circuits rise through the dorsal thalamus into the cerebrum. The dorsal thalamus is enveloped by a sheet of inhibitory neurons. This inhibitory envelope is called the *reticular nucleus of the thalamus* (RNT, RTN), even though it is not nuclear in shape. It should be distinguished from the *reticular formation of the brain stem*, which is involved in the more primitive sensation of pain. The RNT and dorsal thalamus together (sometimes called *paleocortex*) may be seen as the phylogenetic precursor of cerebrum. Both paleocortex and cerebrum exhibit *on-center off-surround* circuitry, a neuroarchitectural design that will become very important in subsequent chapters.

The basal ganglia comprise a group of structures including the *caudate nucleus*, the *putamen*, and the *globus pallidus*. Not visible in figure 4.1, these structures lie behind the plane of the medial section, lateral (alongside) and anterior to (to the front of) the thalamus. They have occasionally been linked to language behavior (Metter et al. 1983; Ullman et al. 1997), but we will view this link largely in terms of their better-documented involvement with posture and gross motor control. The basal ganglia are particularly influenced by dopaminergic signals arising just below in the *substantia nigra* of the midbrain, and deterioration of this dopaminergic pathway is an immediate cause of Parkinson's disease.

## The Limbic System

Ventral to (beneath) and surrounding the thalamus are the various structures of the *limbic system*. The *hypothalamus* (Hy in figure 4.1) is the seat of physiological drives (hunger, sex, fear, aggression). Often classified as a gland, it secretes many hormones, such as vasopressin and oxytocin, and exerts direct influences on the autonomic nervous system, which in turn controls such functions as heart rate and breathing. The hypothalamus also directly affects the brain's *pituitary gland*, which secretes a wide range of other hormones. As noted in chapter 3, these hormones are sometimes classed as neurotransmitters, but since they are diffusely circulated with effects beyond the nervous system, I prefer to class them simply as hormones, reserving the term "neurotransmitter" for chemicals with more cognitive synaptic effects.

Lateral to the hypothalamus is the *amygdala*. The amygdala sends projections directly to the caudate nucleus and is sometimes therefore counted among the basal ganglia. The amygdala is quite directly connected to the frontal and the motor cortex and has been implicated in the modulation of emotions.

The *hippocampus* (Hc in figure 4.1) is the amygdala's sensory counterpart. In 1966, Milner described a patient, HM, who had suffered from severe temporal lobe epilepsy (see also Milner et al. 1968). In a standard effort to control the epilepsy, surgeons removed large areas of temporal cortex. In HM's case, they also removed substantial portions of the hippocampus. HM's epilepsy was brought under control, but a new problem was created. HM developed *anterograde amnesia*: he "forgot the future." HM was unable to form new memories. He was unable to learn. He remembered well his wife, friends, and family and his old neighborhoods and haunts. He could not, however, remember people he met since his surgery. Each day, he would meet with his doctor and say, "Have we met?"

Milner's report caused an instant sensation in the neuroscience community and inspired numerous studies of the hippocampus. The most popular theoretical explanation of hippocampal function has been a consensus "buffer" model (see the four-volume series edited by Isaacson and Pribram for the evolution of this model). In this consensus model, the hippocampus is viewed as "working memory," rather like a computer coprocessor or RAM cache which performs operations on sensory input or briefly stores data before it is transferred to long-term storage in "declarative memory." This model has the attraction of computational metaphor, but it accords a large cognitive role to a small structure in a brain region otherwise not found to be particularly "cognitive." It is also unable to explain many facts. For example, HM was able to develop long-term memory for certain unemotional forms of knowledge like the solution to the Tower of Hanoi problem.[3] More plausible from our perspective is Gray's theory of the hippocampus as a "comparator" (Gray 1975, 1982). Gray's theory also appeared in the Isaacson and Pribram series (Gray and Rawlins 1986), but it represented a minority opinion and has been slower to gain popularity.

Situated posterior to the amygdala and lateral to the hypothalamus, the hippocampus is also widely connected to the cerebrum and the *cingulate gyrus* (sometimes called the "limbic lobe"; (CG in figure 4.1) via the *cingulum*. The cingulum is a massive bundle of axons that originate in the parahippocampal lobe of the temporal lobe and arch up and around the diencephalon, behind the cingulate gyrus in figure 4.1. (In figure 4.10, the left cingulate gyrus has been dissected, exposing the cingulum.) Indeed, the hippocampus is so minimally differentiated from cerebral cortex that it is sometimes called the "hippocampal lobe" of cerebrum. But the hippocampus is also connected to the *mammilary body* of the hypothalamus via the *fornix*. In Gray's theory, the limbic system—and the hippocampus in particular—can be seen as moderating between the cognitive information of the cerebrum and the physiological and emotional drives of the more primitive brain.

We will return to this issue several times in subsequent chapters. For the present, we simply observe how the tears and tantrums of any two-year-old demonstrate that learning can be a very emotional business, and that the case of HM suggests that learning can fail if it is disconnected from primitive survival instincts and drives.

## The Cerebellum

Dorsal to the pons is the *cerebellum* (Cblm in figure 4.1). The cerebellum, like the cerebrum, is composed of a rind, or cortex, wrapped around several deep nuclei (CN in figure 4.2) and divided into two hemispheres. The cerebellum is especially notable for its control and coordination of fine motor behaviors like knitting, playing a musical instrument, or speaking a language. This control does not, however, include the planning or initiation of behavior, so as we shall see, the cerebellum's role is considerably less "cognitive" than the cerebrum's.

The distinctive cerebellar architecture (figure 4.2) is characterized by large output cells, the *Purkinje cells* (PC in figure 4.2), which send signals to cerebellar subcortical nuclei (CN). Purkinje cells are innervated by *climbing fibers* (CF), nonspecific inputs that arise from the inferior olive of the medulla, and by long by *parallel fibers* (PF) arising from *granule cells* (GC), which are the cerebrum's principal input of sensory and motor information. Granule cells are innervated by afferent *mossy fibers* (MF).

The Purkinje cells are embedded in the regular grid of the cerebellum's parallel fibers, forming a matrixlike architecture which is schematized in figure 4.3. The fact that this architecture can be efficiently modeled by using well-known shortcuts like matrix algebra made it easy to quickly program complex, putatively cognitive, computational neural networks. But unlike cerebral pyra-

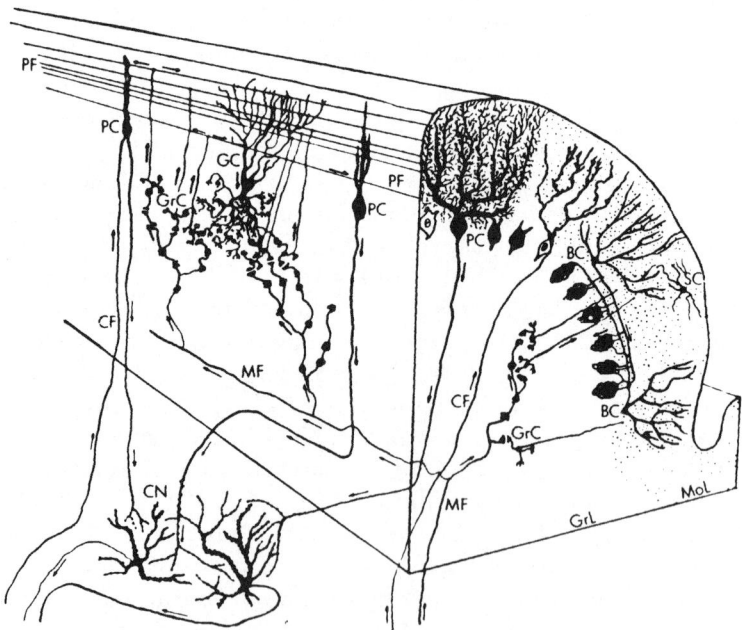

**Figure 4.2.** Cerebellar cortex. (Fox 1962. Reprinted by permission of Appleton and Lange.)

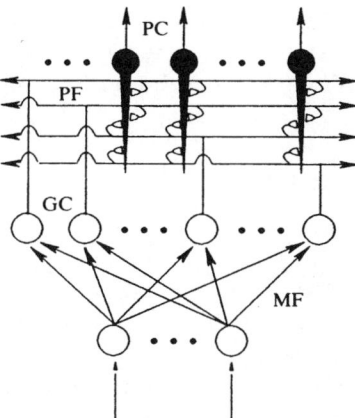

**Figure 4.3.** Schematic of cerebellar cortex. (Loritz 1991. Reprinted by permission of Oxford University Press.)

midal cells, cerebellar Purkinje cells are *inhibitory*. As we shall see, the cerebellum therefore does not have the on-center off-surround architecture of cerebral cortex. As one result, the cerebellum is "swept clean" of residual neural activity within 0.1 ms of input (Eccles 1977). This is all well and good for the processing of fast and fine motor commands, and in later chapters we will see how it is also crucial to the fluent pronunciation of language, but a 0.1 ms short-term memory ill-serves what most scientists would call "cognition." In comparison with cerebral architecture, we will conclude that the cerebellar architecture is fundamentally noncognitive. The important point here is that language is not and cannot be learned by just any brain cells: it is, with several interesting qualifications, really learned only by *cerebral* cells, and so it is to the cerebrum that we turn our major attention.

## The Cerebrum

After the skull is opened, it is the *cerebrum* that first presents itself to the surgeon. On first viewing, the cerebral cortex looks like a thick placemat (or, yes, a thin rind) about 5 mm in thickness, which has been wrinkled and stuffed into a too-small cranium. This is the traditional view of the cerebrum, as well as the one most familiar to the lay reader, and it is the view with which we shall begin.

But there are more illuminating ways to look at the cerebrum. As we have seen, Ramón y Cajal used the microscope to take a quite different view. He put the cerebral sheet on edge and studied it from the perspective of its thickness. It was in this view that the cytoarchitecture of cerebral cortex was first exposed, and this will be the second perspective from which we shall view cerebral cortex.

A third view of the cerebrum combines the surgical and cytoarchitectural views. This is the "planar" view of the cerebrum, and it is the view that will be most important to our subsequent development of adaptive grammar.

The surgical view of the cerebrum:
the cerebral hemispheres

The human cerebrum is fraught with folds and bulges. These (usually larger) *fissures* and *lobes* or (usually smaller) *sulci* and *gyri* were the basis of the earliest anatomical attempts to describe the brain by its shape. No two brains are wrinkled in exactly the same way, but the larger folds and bulges are common to all human brains. Later, the finding that these common lobes often process specific types of information gave rise to the computational metaphor of a "modular" brain.

Like most chordate anatomies, the cerebrum (as well as the subcerebral brain) is bilaterally organized and divided into left and right hemispheres by a deep central sulcus. As originally suggested in figure 2.9, the right hemisphere moves and senses the left body while the left hemisphere moves and senses the right body, and as noted in chapter 1, language is primarily processed in the left hemisphere. The right hemisphere is more "lateralized for" spatial tasks. In other major respects, however, the hemispheres are essentially identical.

Figure 4.4 shows the major fissures of the brain, as well as cytoarchitecturally distinguishable areas of cortex known as Brodmann areas. Shown are a lateral view (an outside-in view from the side) of the left hemisphere and a sagittal (an outside-in side view—not a dead-center, medial view) of the left hemisphere.

Mapping motor cortex

Each cerebral hemisphere is divided by a long, vertical *fissure of Rolando*, the S-shaped line running down the center of figure 4.4 (top). The *frontal lobe*, which lies anterior to this central sulcus, plans actions and so is often called *motor cortex*.

In the 1950s through the 1970s, before the development of radiological techniques, surgical removal was the most effective treatment for brain tumors. Since the brain itself has no neurons which can directly sense pain, brain surgery only requires a local, scalp anesthetic, and after the skull has been opened, the patient can remain conscious during the procedure. This requires great courage on the part of the patient, but it is also a great contribution to the safety of the procedure. With the brain exposed, the surgeon can stimulate different regions of the brain with an electric probe while the patient reports what he senses. Using this information, the surgeon can avoid accidentally damaging speech areas or other especially critical areas of the brain. The composite of a number of such surveys revealed that many other primary motor and sensory functions are localized in the same way that language is localized in Broca's and Wernicke's areas.

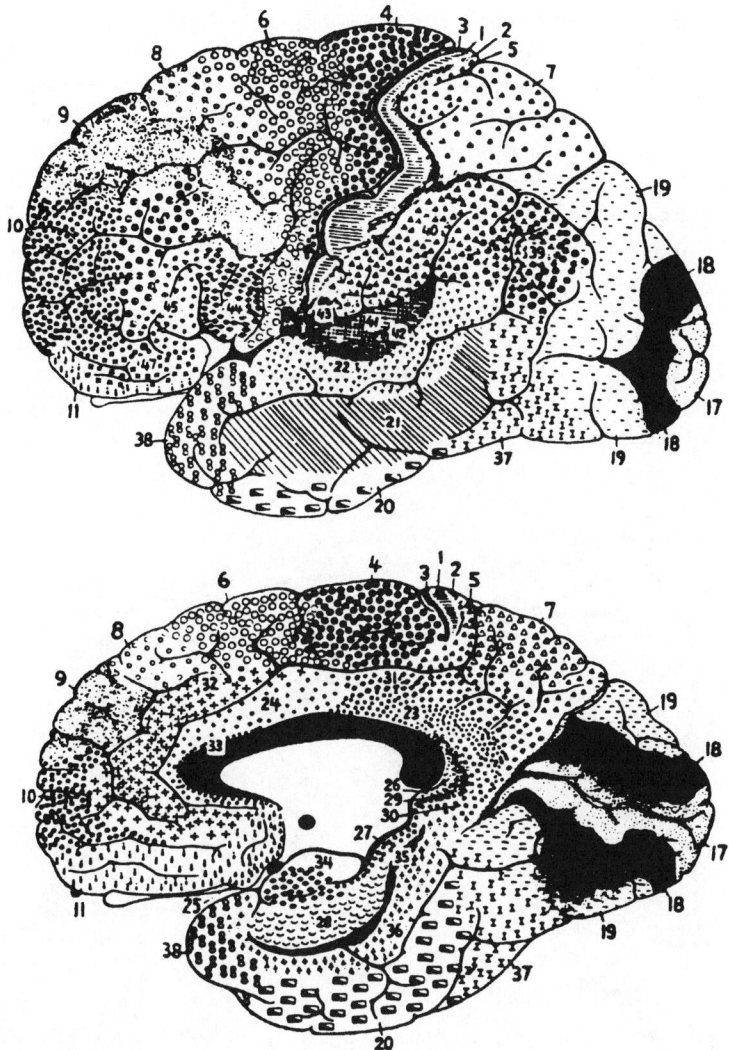

**Figure 4.4.** Brodmann areas. Lateral (top) and medial (bottom) views of the left cerebral hemisphere. (Brodmann 1909. Reprinted by permission of J. A. Barth Verlag.)

Before the discovery of genetics, it was believed that human sperm cells contained a *homunculus*, a little man, which grew up into a bigger baby. This belief is now regarded as a preposterous example of medieval pseudoscience, so it was quite surprising when Wilder Penfield's preoperative surveys discovered just such a homunculus in the brain! Penfield showed that the primary motor area, which is the gyrus just anterior to the fissure of Rolando (Brodmann areas 4 and 6 in figure 4.4), sends signals to body parts as if a little homunculus were laid upside down on the gyrus (figure 4.5; Penfield and Rasmussen 1950).

In figure 4.5, mouth, lips, and tongue of the motor homunculus are drawn disproportionately large because they are controlled by disproportionately large areas of motor cortex. Apparently, eating well is as important to humans as it was to early coelenterates. But when the human homunculus's mouth and lips are compared with similar studies of apes and other animals, we find they are still comparatively very large. For humans, speaking well may be even more important than eating well: the large area of motor cortex that the human motor homunculus devotes to mouth and lips must correspond to the complex motor planning required by human speech. And indeed, it is in this frontal region that Broca's area is located, just anterior to the primary cortex for mouth, lips, and tongue. In fact, the entire forebrain extending frontward beyond Broca's area is comparatively greatly enlarged in *Homo loquens*. In fig-

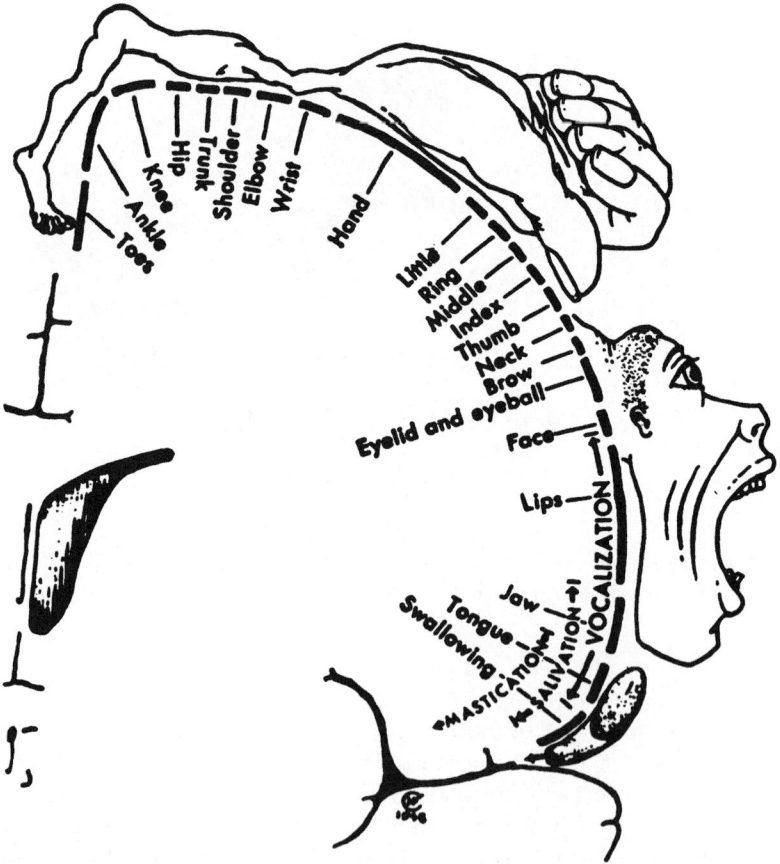

**Figure 4.5.** The motor homunculus. (Penfield and Rasmussen 1952. Reprinted by permission of Simon and Schuster.)

ure 4.4 (top) Broca's area is in the vicinity of Brodmann areas 44 and 45. Because Broca's area is located in motor cortex, aphasics with damage to Broca's area are often called *motor aphasics*.

Sensory cortex and touch

Neocortex in the gyrus just posterior to the central sulcus responds first to sensation and is often called *primary sensory cortex*. In this gyrus lies a second, twin homunculus (figure 4.6). This one maps the receptive, sensory regions of the cerebrum.

This somatosensory homunculus receives afferent nervous signals from touch receptors throughout the body. But touch is of secondary interest to our investigation of cognition. Vision and hearing are much more important—so important in the chordate brain that they occupy not just parts of a small homunculus but whole lobes of the brain.

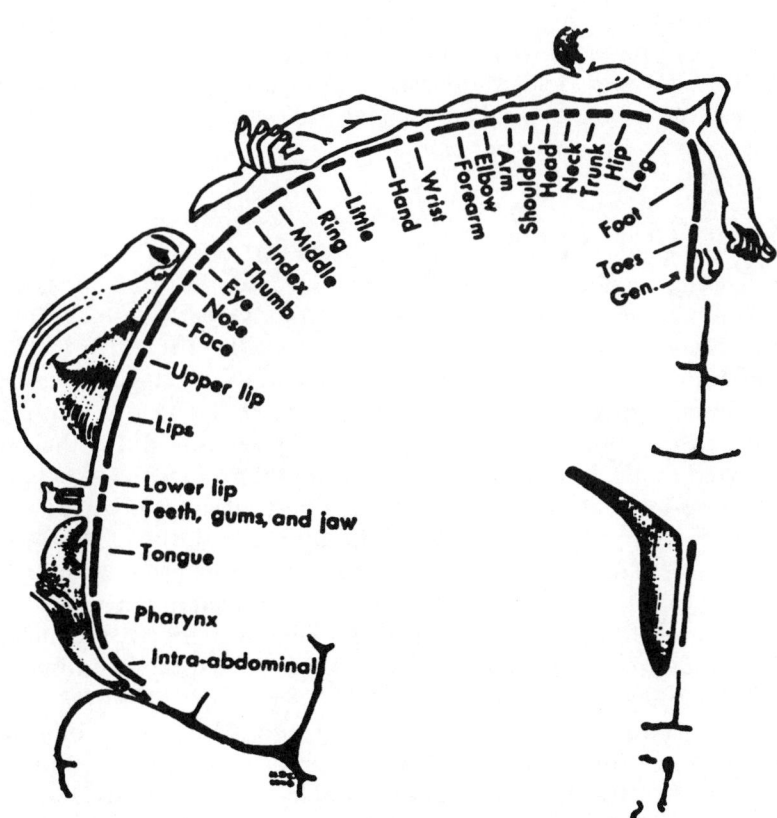

**Figure 4.6.** Somatosensory homunculus. (Penfield and Rasmussen 1952. Reprinted by permission of Simon and Schuster.)

### The temporal lobe: auditory cortex

A second large sulcus, the *Sylvian fissure*, divides each hemisphere horizontally. In figure 4.4 (bottom), this is the black estuary running from seven o'clock toward two o'clock. Ventral to this fissure and beneath the skull bone for which it is named lies the *temporal lobe* (see also figure 4.1). The auditory nerve erupts into cortex here, within the fissure, in Brodmann area 34. This area is variously identified as *koniocortex, Heschl's gyrus*, or *primary auditory cortex* (see also figure 12.1). Wernicke's area is found posterior to koniocortex, in the posterior part of the superior temporal gyrus, in the vicinity of Brodmann areas 41, 42, and 22 (figure 4.4, top).

### The parietal lobe: association cortex

The parietal lobe is situated above the Sylvian fissure and behind the fissure of Rolando. It, too, is named for the skull bone above it. The parietal lobe is visibly distinct from the occipital lobe behind it only when the cerebrum is viewed from within (as in figure 4.4, bottom, or figure 4.1). As noted above, the anterior gyrus of the parietal lobe is *primary sensory cortex*, specifically dedicated to reception of primary sensory inputs from touch. The posterior parts of the parietal lobe, however, yield only diffuse responses to tactile stimulation or focal lesions. In consequence, parietal cortex is called *association cortex* because it diffusely associates primary percepts.

### The occipital lobe and vision

The *occipital lobe*, again named for the skull bone under which it lies, is the most caudal (toward the tail) lobe of the cerebrum. It is clearly defined only when the brain is viewed from beneath or from within, where it is clearly delimited by the parieto-occipital fissure (as in figures 4.1 and 4.4). The main visual pathways terminate here in *primary visual cortex*, or *striate cortex*, after running from the retina through the thalamic LGN. It at first seems odd that the occipital lobe, located at the very back of the head, should process the visual percepts from the eyes in front, but when we recall that frogs and other prey animals have eyes where humans have ears, the better to watch their backs, the location is not so strange. Indeed, we can learn much by close examination of animals. So even though this book is mostly interested in language, much of what we know about complex neural processing is derived from studies of vision in lower animals, and we must devote some space to a discussion of the vast scientific literature on vision.

In 1967, Hartline and Granit received the Nobel Prize for their work on excitation and inhibition in animal visual systems (e.g., Hartline and Graham 1932; Granit 1948; Hartline 1949; Hartline and Ratliff 1954; see also Ratliff 1965). In particular, Hartline studied the connection patterns in the retina of the "primitive" species *Limulus polyphemus*, the horseshoe crab. Hartline found that the cells of the crab's eye were arranged in an *off-center off-surround* anatomy (figure 4.7).

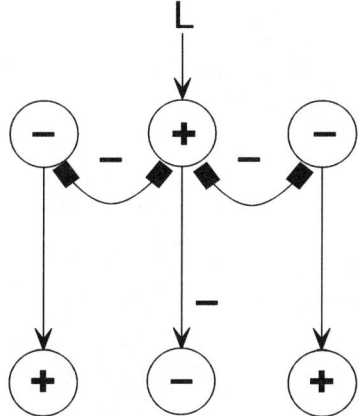

**Figure 4.7.** An off-center off-surround anatomy.

This anatomy suggests that *Limulus* perceives its world like a photographic negative. That is, when light strikes a cell in the *Limulus* retina (L in figure 4.7) that cell inhibits its corresponding postretinal cell, creating a black percept in response to light. But the cell also inhibits its surrounding retinal cells. These cells, being inhibited, no longer inhibit their corresponding postretinal cells. Being thus disinhibited, the postretinal cells become active, creating an "on" percept in response to light stimulation being "off."

We shouldn't be *too* sure that *Limulus* perceives black as white, however. Just as $(-1)(-1) = +1$, inhibiting an inhibitory neuron can lead to the excitation of another neuron. This kind of sign reversal can confuse our attempts to simply relate neurotransmitters to behavior. So, for example, adrenaline has an excitatory effect upon behavior, but in the brain, as the neurotransmitter *noradrenaline*, it has been found to have an inhibitory effect, hyperpolarizing the postsynaptic cell membrane (see figure 3.10).[4]

In mammalian vision, similar networks occur, but they are *on-center off-surround* networks. This *on-center* off-surround anatomy is a significant evolutionary advance. In on-center off-surround anatomies like that in figure 4.8, some center cells, say in the LGN of the thalamus, are excited by a stimulus. Those cells relay information up to cells in striate cortex. These, in turn, echo excitation back to the center cells, keeping them on. In the cerebrum (figure 4.8, top), afferent axons branch into the dendritic arbor of pyramidal cells (in the background) and a large basket cell (in the foreground). The basket cell sends axon collaterals to surrounding pyramidal cells, inhibiting them. At the same time, similar inhibitory cells in the thalamic reticular formation (in figure 4.8, bottom) inhibit surrounding thalamic relay cells.

In contrast to the black-is-white world of *Limulus*, we can think of the on-center off-surround anatomy as creating a world of sensation in which white is white and black is black. This does not confer any particular evolutionary visual advantage (*Limulus* has been around for a long, long time), but on-center

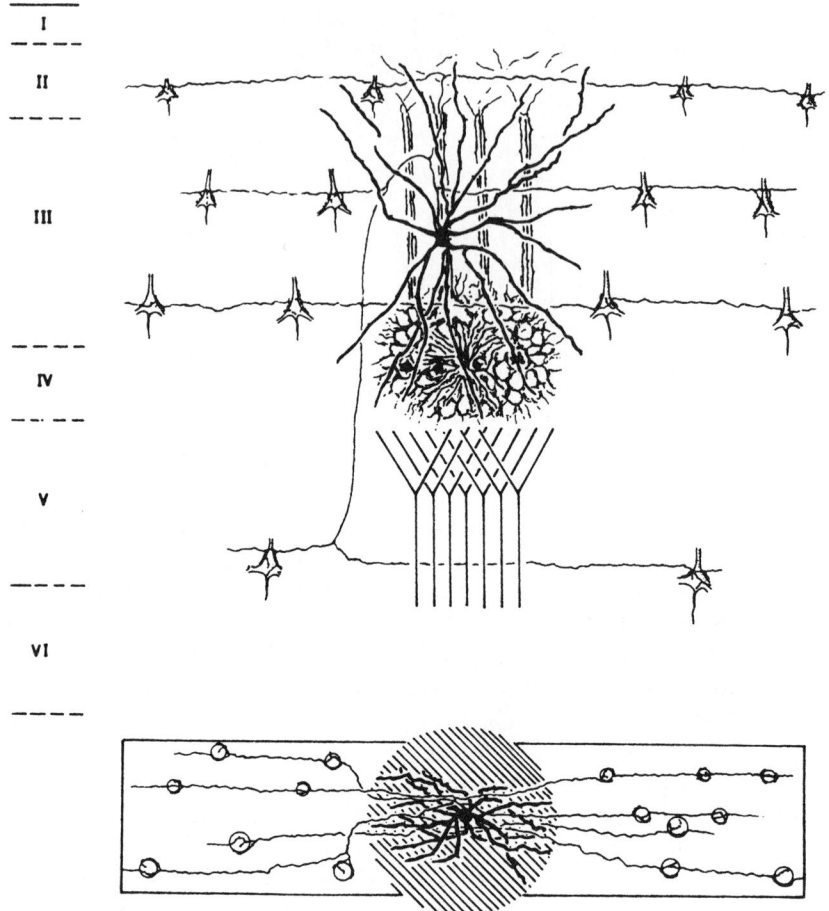

**Figure 4.8.** A thalamocortical on-center off-surround anatomy. (Jones 1981. Reprinted by permission of MIT Press.)

off-surround anatomies are not restricted to vision. They are found at all levels of anatomy from our six-celled brain in chapter 2 and the brain stem on up to the thalamic reticular formation and the cerebrum, where they are ubiquitous. As we shall see, what especially confers an evolutionary advantage on the on-center off-surround anatomy is that it can *learn*, something for which *Limulus* is not renowned.

"Modules" and intermodular connections

The lateralization of language to the left hemisphere, the localization of speech in Broca's area and speech understanding in Wernicke's area, and the presence of homunculi in primary sensory and motor cortex suggested that there are places in the brain for different processes, just as there are places or *mod-*

*ules* in a computer program for different subprograms. Perhaps the most spectacular evidence for such modularity has come from Sperry's work.

The underside of the cerebral cortex is covered with tendonlike "white matter." Until the discovery of nerve cells, such white matter was thought to be simply a ligature, holding the lobes and hemispheres of the cerebrum together. Now we know that the white matter actually consists of bundles of pyramidal cell axons, neural highways across which brain cells communicate. In patients suffering from severe epilepsy, Sperry bisected the *corpus callosum*, the massive bundle of nerve fibers connecting the right and left cerebral hemispheres (Sperry 1964, 1970a, 1970b, 1967; see figure 4.10 and CC in figure 4.1). Sperry's reasoning was that since massive, grand mal seizures exhibited nervous signals reverberating out of control back and forth between the hemispheres, severing the corpus callosum would stop this pathological resonance. The procedure was dramatically successful, but almost equally dramatic were some of the patients' postoperative sequelae.

For example, when such a "commisurectomized" patient was blindfolded and an apple placed in her left hand, she salivated and otherwise recognized that the object was food, but she could not *say* she was holding an apple. This happened because the sense of touch projects contralaterally from the left body to the right brain and from the right body to the left brain. As Broca noted, most people's language is lateralized to the left hemisphere. So the patient's left hand sent sensory touch signals up the brain stem to her right hemisphere, where the apple was behaviorally recognized as food, but since the corpus callosum had been cut, the right hemisphere could not relay this information to the language modules of the left hemisphere, and the patient could not say she was holding an apple. Fortunately for commisurectomized patients, the cognitively important senses of vision and hearing are not as strongly lateralized as touch.

Vision is strictly contralateral in that the left visual field projects to the right hemisphere, and vice versa, but both eyes have both a left and a right visual field, so as long as the commisurectomized patient has both eyes open, he can see normally. Hearing is also less lateralized in the sense that each ear sends not only contralateral signals but also *ipsilateral* (same-side) signals. As it happens, the contralateral connections are stronger, allowing a dichotic listening test (Kimura 1967) to identify the language-dominant hemisphere even in normal subjects without localizable brain lesions. In a dichotic listening test, minimally contrasting words like *bat* and *pat* (*minimal pairs*) are simultaneously presented in stereo, one to each ear. Asked what word they hear, most subjects will most often report the word presented to the right ear. This indicates that the contralateral left hemisphere is dominant for language.

More recently, another region of "intermodular" interaction has been identified in the *angular gyrus*, which lies at the intersection of the parietal, temporal, and occipital lobes, in the vicinity of Brodmann area 40 (figure 4.4, top). Focal lesions to this area of the cerebrum have resulted in rather pure *alexia*—the inability to read. This observation suggests that the modules for hearing, language, and vision send their outputs to the angular gyrus for processing

66 • HOW THE BRAIN EVOLVED LANGUAGE

during reading. The angular gyrus lies on a diffuse intermodular pathway connecting all of these modules, the *arcuate fasciculus*.

The arcuate fasciculus

Wernicke, noting the functional correlation of language understanding with the brain region that bears his name, was the first to propose the modular theory of brain structure. In the Wernicke-Lichtheim model (figure 4.9), a lesion at 1 disrupts input from the ear to Wernicke's area and corresponds to hearing impairment or deafness. A lesion at 2, pure Wernicke's aphasia, allows sounds to be heard, but word perception and recognition are impaired. A lesion at 3, receptive aphasia, allows words to be recognized, but the comprehension of speech by association cortex is impaired.

On the productive side, a lesion at 6 disrupts output from Broca's area to the mouth. This corresponds to *dysarthria*, the motoric (as opposed to cognitive/aphasic) inability to articulate speech. A lesion at 5, pure Broca's aphasia, allows speech sounds to be uttered, but word production is impaired. A lesion at 4, productive aphasia, allows words to be uttered and repeated, but connected speech is impaired.[5]

The *arcuate fasciculus* (figure 4.10; 7 in figure 4.9) connects Broca's area and Wernicke's area. Wernicke predicted that a lesion to the arcuate fasciculus would produce a *conduction aphasia* that would present the quite specific inability to perform *verbatim* repetition. The conduction aphasic would be able to understand speech because the pathway 1-2-3-4-5-6 would remain intact. He would be able to produce speech because the same pathway would also be intact in the opposite direction. Following these pathways, he would be able to accurately paraphrase what is said, but he would be unable to repeat language

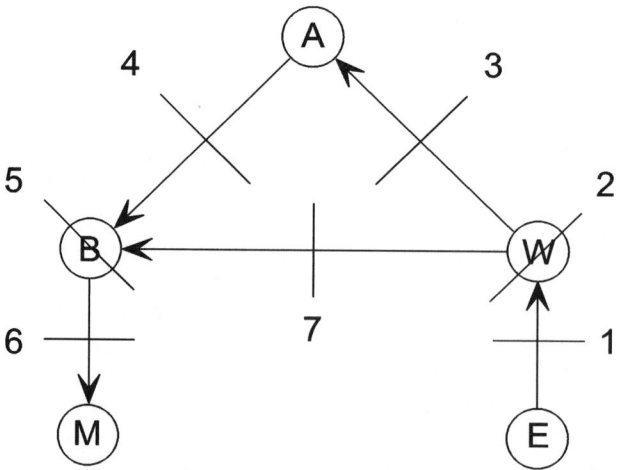

**Figure 4.9.** The Wernicke-Lichtheim model of language cortex. (After Lichtheim 1885.)

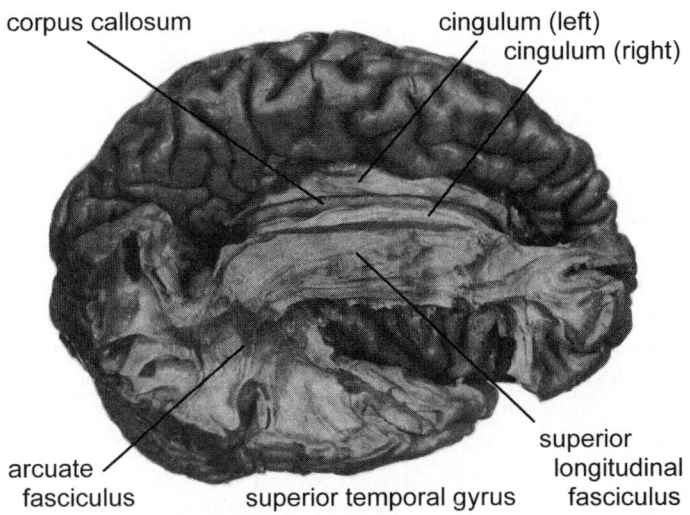

**Figure 4.10.** The arcuate fasciculus, right hemishphere. (Montemurro and Bruni 1981. Reprinted by permission of D. G. Montemurro.)

verbatim along pathway 1-7-6! Although only a few cases of conduction aphasia have been unambiguously diagnosed (typically resulting from deep brain tumors), they have proved Wernicke right: Broca's area and Wernicke's area *do* exchange information across the arcuate fasciculus, via the angular gyrus.

Broca's work and Wernicke's work inspired a century of research into the localization of language and other brain functions and led to the first scientific understanding and treatments of aphasia, epilepsy, and other brain and language disorders. By the 1950s, however, psychologists and physiologists had begun to question localization (Lashley, "In Search of the Engram," 1950). Under the rising influence of "modular programming" in computer science (Wirth 1971; Parnas 1972), research into localization persisted under computational metaphor, but Lashley's "engram" was not to be found in any one place. Language "modules" like Broca's area and Wernicke's area were indistinct and variable. Indeed, there was even found to exist a small population of otherwise normal people whose language is localized in the *right* hemisphere. Nor was this simply an isolated anomaly: a much larger minority (most left-handers) were found to have language fairly evenly distributed across *both* hemispheres. Then there were cases of aphasic children. For adult aphasics, the prognosis is bleak. Once the language hemisphere is damaged, recovery is usually incomplete and often minimal. But for child aphasics, the prognosis is miraculously good. Within several years, the other hemisphere or a spared gyrus takes over the language functions of the damaged region, and recovery is often—even usually—complete.

There was also the case of sign language, mentioned in chapter 2. As it became accepted that the sign languages of the deaf were cognitively complete, it also became apparent that they are initially processed by manual cortex and visual cortex, not the oral and auditory cortex proximate to Broca's module and Wernicke's module. Although some recent fMRI (functional magnetic resonance imaging) studies of deaf signers do show some elevated activation in Broca's area and Wernicke's areas during signing, the activation is relatively small compared with spoken language (Neville and Bavelier 1996). A reasonable explanation is that these are resonances created by the existence of the arcuate fasciculus rather than vestigial activation of a local "module."

It can be said that vision is innately hardwired to striate cortex in the occipital lobe, for this is where the optic nerve erupts into the cerebrum. It can be said that hearing is innately hardwired to koniocortex in the temporal lobe, for this is where the auditory nerve erupts into the cerebrum. It can be said that touch (as in Braille reading) is innately hardwired to the tactile-sensory regions of the parietal lobe, for this is where tactile nerves erupt into the cerebrum. It can be said that speech articulation is innately hardwired to posterior Broca's area in the motor regions of the frontal lobe, for this is where signals from motor pyramidal cells leave the cerebrum, bound for the articulators. But it seems hard to say that language itself, like a sixth sense, is similarly hardwired to any particular place in the cerebrum. The cerebrum is plastic. (Indeed, Roe et al. [1990] successfully surgically rewired a monkey's optic nerve to auditory cortex!) As recovered child aphasics show, language can be almost anywhere in the cerebrum.

If there is a "module" for language, then the evidence we have seen would suggest that the module is neither Broca's area nor Wernicke's area nor the angular gyrus, nor even anything specific to the left hemisphere. If there is a "module" for language, it would seem that the best physiological candidate is the arcuate fasciculus, but the arcuate fasciculus is neither language specific nor really a "place" in the brain. Science always progresses by ex-pressing the unknown with metaphors of the known, but the localist, computational metaphor of "modularity" seems exhausted. Like calling the telephone company "the building across the street," metaphor can wrap new science in terms which the mass of researchers will find familiar and meaningful (Kuhn 1962), but familiar metaphor can also obscure what is novel, distinctive, and essential.

The laminar structure of neocortex

Thus far, we have been concerned primarily with the cerebrum as it appears to the unaided eye, a wrinkled rind covered with bone. But if we were to take the cerebral cortex out of its skull, snip its white-matter ligaments, and then unfold it, we would get a rather different picture. We would find that the cerebrum is actually a 0.5 $m^2$ sheet of tissue about 4–5 mm thick. Instead of a distinctive geography of folds and bulges, we would see only an undifferentiated plain of gray matter above and white matter below. To find structure, we would need to look inside the sheet with a microscope.

Anatomical studies (see figures 3.1 and 4.8) count six or seven cytologically distinct laminae in cerebral cortex. We, however, will focus on only three.

In most areas of cortex, we can distinguish three layers in which (stained) pyramidal cell bodies predominate. In figure 3.1, these are in laminae III, V, and VI. Anatomical studies number these laminae in the order in which the anatomist encounters them, from the outside in. Ontogenetically, however, the first pyramidal cells to develop are those in the lowest layer, so unless I specifically use roman numerals in citing the anatomical literature, these are the cells I shall call "first." These first pyramidal cells migrate out from the embryonic brain to form a first layer of cortex. Phylogenetically, it is as if a mutation occurred in chordate evolution, causing not one but two thalamic reticular formations to develop, the second (neocortex) enveloping the first (paleocortex) and becoming the cerebrum. Later, a second and a third self-similar layer of on-center off-surround brain developed, each enveloping the former.

It is reasonable to assume that the first layer of early-evolving and early-developing pyramidal cells are largest because they are oldest. As the largest, they have large dendritic trees which rise and branch high above their cell bodies. Their myelinated axons project far and wide below the cortical sheet. Many of these axons, notably those in the primary motor strip (the "motor homunculus," figure 4.5), project far down into the basal ganglia and beyond. These and many others also send widely branching axon collaterals, forming the corpus callosum, the arcuate fasciculus, and the entire web of white matter beneath the cerebral sheet. Eventually these collaterals rise up again elsewhere into the neocortical sheet, to innervate other neurons far from their originating cell bodies.

Later in ontogeny, smaller pyramidal cells develop. They migrate outward from the embryonic brain, past the first layer of pyramidal cells, to form the second layer. In most of cortex, these are the primary sensory cells of neocortex. Afferent pathways, rising up from the sensory organs through the thalamus, synapse preferentially upon these cells. Like the older cells of the first layer, their myelinated axons also project into the neocortical white matter, but they project more locally into neighboring neocortex, or they return reciprocal signals back to the thalamus.

Finally in ontogeny, a third cohort of pyramidal cells migrates to form a superficial layer near the top of the cortical sheet. These are the smallest pyramidal cells. Their axons and dendritic trees are small, and their axons are the most local of all.

This pattern is repeated everywhere in cerebral cortex with one major thematic variation: wherever cerebral input is concentrated, the second layer is especially densely populated with especially large pyramidal cells, and wherever cerebral output is concentrated, the oldest layer is especially densely populated with especially large pyramidal cells. Thus, in striate cortex, where the optic nerve pathway enters neocortex, and in koniocortex, where the auditory nerve pathway enters neocortex, the middle layer is densely populated with largish pyramidal cells. Similarly, in the bottom pyramidal layer of primary motor cortex, where motor signals exit cortex, the pyramidal cells are relatively large and dense.

## Neurogenesis

One of the great mysteries of the brain has been how a trillion-odd cells manage to wire themselves together in anything like an orderly fashion. How is the brain born? How does it develop? It has long been clear that neurons atrophy and die in the absence of nervous stimulation. Autopsies of the spines and brains of amputees reveal neuron atrophy and degeneration extending several synapses away from the amputation. In animal studies, this natural experiment has been refined to scientific technique, and much of what we presently know about the connectivity of nerve pathways has been learned by oblating sections of animal nervous systems and following the resulting patterns of atrophy and degeneration.

Such facts, coupled with the discovery of chemical neurotransmission accompanying neural stimulation, led to the general theory that neurons depend for their very life upon the kind of neural import-export policy proposed at the end of chapter 3. But if (excitatory) neurotransmitters of the sort Dale discovered in the 1920s were the currency in this neural exchange, it was not until the 1980s that the goods were identified. In 1986, Levi-Montalcini and Cohen received the Nobel Prize for their discovery of nerve growth factor (NGF).

Although NGF and related neurotrophins first appeared to simply be a new but important class of growth-stimulating hormones, it has recently come to be understood (Thoenen 1995) that their synthesis and efficacy are critically dependent upon neural activity. Simultaneously, nitric oxide (NO) was discovered to be a ubiquitous "retrograde messenger" that is released from the activated postsynaptic cell and taken up by the stimulus-activated presynaptic axon terminal, there to facilitate growth and the production of more neurotransmitter (Kandel and Hawkins 1992). The details are still the subject of cutting-edge research, but from these pieces we can begin to develop a picture of how the brain is built. This emergent picture is very much one of "neural Darwinism" (Edelman 1987), in which the developing brain not only grows but also evolves in a complex, neuroecological interplay of competitive and cooperative responses to the environment.

## Columnar Organization

In Golgi-stained sections (figure 3.1), the apical dendrites of cerebral pyramidal cells stand out as pillars of neural structure, and close examination has revealed that inputs to and outputs from cerebral cortex are all perpendicular to the neocortical sheet. (Contrast this with the parallel fibers of the cerebellum in figure 4.2.) From such observations arose the "columnar model" of cerebral organization.

In the columnar model, the functional unit of cerebral processing is taken to be a multicellular column (Szentágothai 1969; figure 4.11). At the center of each such column, we imagine a large pyramidal cell. Specific afferent inputs

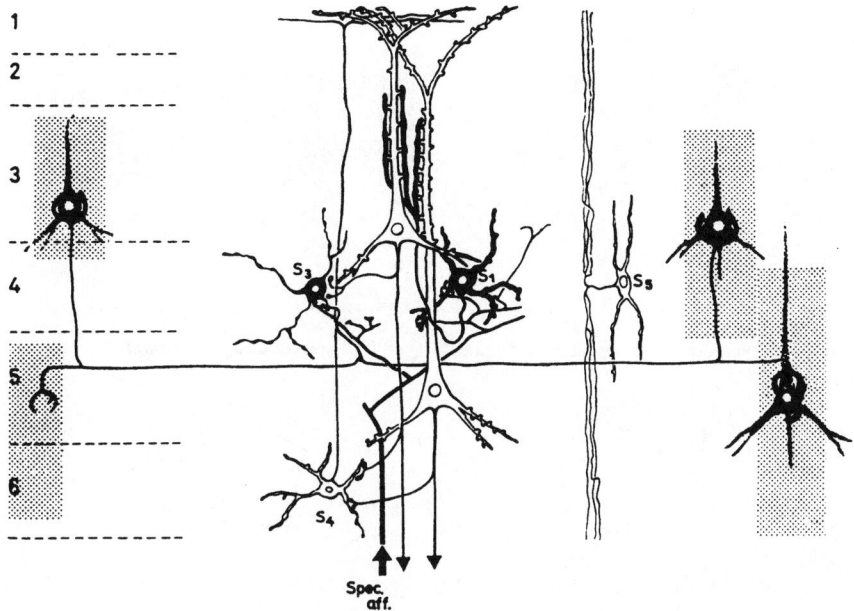

**Figure 4.11.** Columnar organization in neocortex. (Eccles 1977, after Szentágothai 1969. Reprinted by permission of McGraw-Hill Book Company.)

(*Spec. aff.* on figure 4.11) arise into the neocortical sheet, innervating smallish stellate cells ($S_n$), and defining a column. One supposes that it is difficult for a small synaptic input from the afferent axon collateral of a distant neuron to trigger a response in a single large pyramidal cell. Rather, the small input triggers a chain reaction among the smaller stellate cells. This chorus then excites the large pyramidal cell of the column. When a column becomes thus innervated, the pyramidal cell eventually reaches threshold and generates the column's output: a volley of spikes is sent along the axon to many other distant columns.

Szentágothai's schematic of this organization (figure 4.11) was developed after experiments by Mountcastle (1957) which demonstrated that neocortex responded better to electrodes placed perpendicular to the cortical sheet (line *P* in figure 4.12) than to electrodes inserted obliquely (*O* in figure 4.12).

Independently of the work of Mountcastle and Szentágothai, Hubel and Wiesel popularized use of the term "column" in another sense (which we will encounter in chapter 5, figure 5.4f), so researchers began instead to use the term "barrel" to refer to the columns of figure 4.11 (Welker et al. 1996). In this metaphor, we think of a single, afferent axon as defining the center of a neural barrel. Within the barrel, a number of excitatory pyramidal and stellate cells become activated by the input, as well as some number of basket (large dark cells in Szentágothai's drawing) and chandelier cells (absent in Szentágothai's drawing), which inhibit surrounding barrels.[6] In this view, the barrel is more of a statistical entity, a kind of distribution of the probability of an afferent axon innervating excitatory and inhibitory cells.

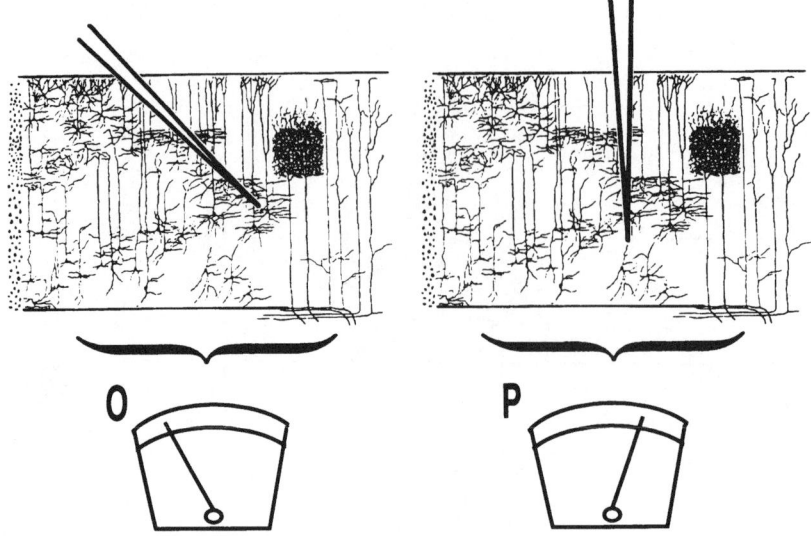

**Figure 4.12.** Perpendicular, not oblique, stimuli activate neocortex.

In either view, we pause to ask what stops the inhibitory cells "in the barrel" from inhibiting the excitatory cells in the barrel. That is, what stops the barrel from committing neural suicide? The answer lies in inspection of the lateral extent of the axon collaterals of the inhibitory cells (figure 4.8). If we stipulate that these collaterals cannot consummate the act of synapsing until they reach a kind of neural puberty, then they can be prevented from synapsing with pyramidal cells in their own barrel. This leads directly to the "planar" view of cortex.

## Planar Organization

In the planar view of cortex, we look down upon the cortical sheet as in the surgical view, but we look more closely. Each afferent input defines the on-center of a barrel, and surrounding that on-center are two concentric rings. Like a pebble dropped in a still pool, there is an on-center peak at the point of impact, and waves ripple out from it. The innermost, inhibitory wave follows a Gaussian probability distribution: it peaks at the radius where axons of most of the barrel's inhibitory cells "reached puberty" and began to form synapses.[7] The outermost, excitatory wave follows a Gaussian probability distribution that peaks at the radius where the barrel's excitatory cells reached puberty. These waves do not simply spread and dissipate, however. They interact in complex patterns with the waves of other barrels.

One of the first researchers to take this planar view and explore these complex patterns was von der Malsburg (1973). Using a variant of the modeling equations developed by Grossberg (1972a), von der Malsburg constructed a pla-

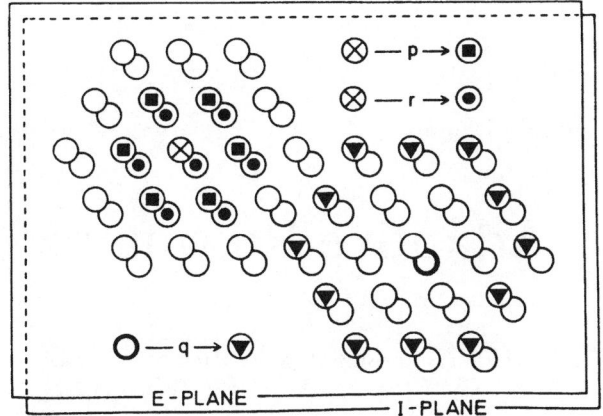

**Figure 4.13.** Von der Malsburg's planar cortex. (Von der Malsburg 1973. Reprinted by permission of Springer-Verlag.)

nar computer model of striate cortex (figure 4.13). Von der Malsburg's simulation used an on-center off-surround architecture to recognize inputs. Early neural network models simply sent the excitatory output of some individual "neurode" directly to some other neurode. Von der Malsburg essentially added the off-surround, inhibitory cells that were missing in figure 4.11. When stimulated, each barrel now increased its own activity *and* decreased that of its neighbor.

Missing, however, from von der Malsburg's model was the fact that in neocortex, barrels also send long-distance, excitatory pyramidal cell output to many other barrels. They also receive reciprocal excitatory feedback from those other barrels. In the next chapter we will build and test a neocortical model that adds these missing elements to the planar model.

• F I V E •

# Adaptive Resonance

> One cannot step into the same river twice.
> Heraclitus

In chapters 3 and 4, we glimpsed the marvelous biochemical and anatomical complexity of the human brain. But in a single breath of a summer wind, a million leaves turn and change color in a single glance. The mind need not read meaning into every turning leaf of nature, but neither the hundreds of neurochemical messengers of chapter 3 nor the forty-odd Brodmann areas of chapter 4 can begin to tally the infinite complexity of an ever-changing environment. To gain even the smallest evolutionary advantage in the vastness of nature, a brain must combinatorially compute thousands and millions of patterns from millions and billions of neurons. In the case of *Homo loquens*, as we estimated in chapter 1, the competitive brain must be capable of computing something on the order of $10^{7,111,111}$ patterns.

But how can we begin to understand a brain with $10^{7,111,111}$ possible configurations? As the reader by now suspects, our technique will be to study minimal anatomies—primitive combinations of small numbers of synapses. First we will model the behavior of these minimal anatomies. Then we will see how, grown to larger but self-similar scales, they can explain thought and language.

We have already seen several minimal anatomies. In chapter 2 we somewhat fancifully evolved a bilaterally symmetrical protochordate with a six-celled brain. Then, in chapter 4, we touched upon Hartline's work detailing the horseshoe crab's off-center off-surround retina and sketched a preview of the on-center off-surround anatomy of the cerebrum. Learning by on-center off-surround anatomies has been the focus of Grossberg's *adaptive resonance theory* (ART), and it is from this theory that we now begin our approach to language.

## From Neocortex to Diagram: Resonant On-Center Off-Surround Anatomies

Figure 5.1a is a reasonably faithful laminar diagram of neocortex, but for simplicity each barrel is modeled by a single excitatory pyramidal cell and a single inhibitory cell. Afferent inputs arise from the white matter beneath the cortex and innervate the barrels. A single fine afferent axon collateral cannot by itself depolarize and fire a large pyramidal cell. So figure 5.1a has the afferent fiber fire smaller, stellate cells first. These stellate cells then fire a few more stellate cells, which each innervate a few more stellate cells, and so on. Eventually, by this kind of nonlinear mass action, an activated subnetwork of stellate cells fires the barrel's large pyramidal and inhibitory cells. The on-center pyramidal cell sends long-distance outputs, while the inhibitory cell creates an off-surround.

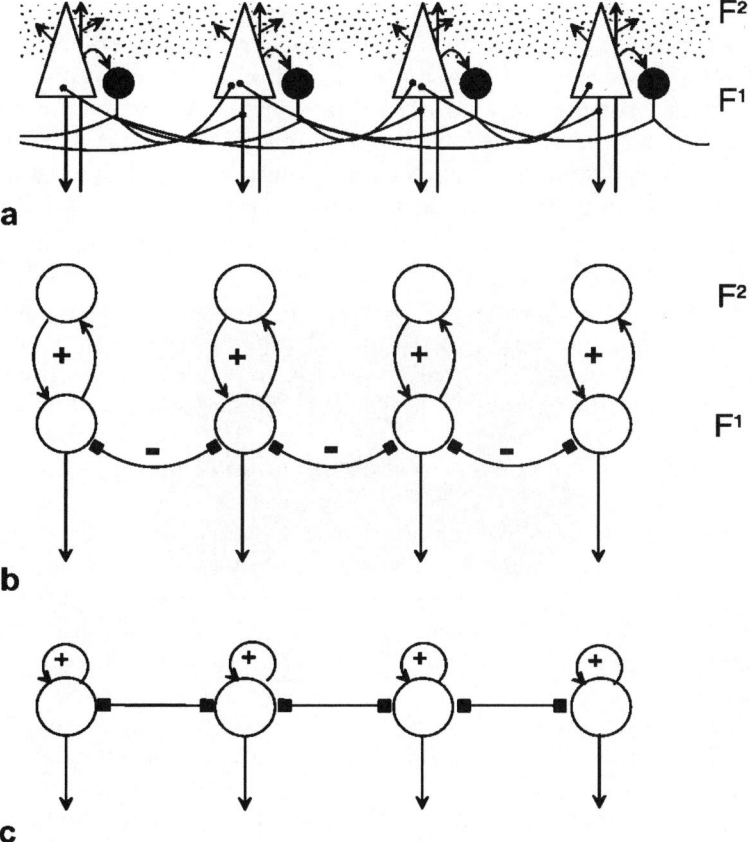

**Figure 5.1.** Three schematics of on-center off-surround anatomies. (a) is a biologically faithful schematic detailing pyramidal cells and inhibitory basket cells. (b) and (c) abstract essential design elements.

In figure 5.1b, we abstract away from Figure 5.1a, and we no longer explicitly diagram inhibitory cells. Following White 1989, we also treat stellate cells as small pyramidal cells, so each node in $F^2$ of figure 5.1b can be interpreted as either a local subnetwork of stellate-pyramidal cells or a distal network of pyramidal cells. In either case, $F^1$ remains an on-center off-surround minimal anatomy.

Figure 5.1c abstracts still further, no longer explicitly diagramming the on-center resonance of $F^2$ nodes. In the diagrams of minimal anatomies that follow, it is important that the reader understand that a circle can stand for one cell or many, while "on-center loops" like those in figure 5.1c can represent entire, undiagrammed fields of neurons. Since we will focus almost exclusively on cerebral anatomies, and since the on-center off-surround anatomy is ubiquitous in neocortex, I will often omit even the on-center loops and off-surround axons from the diagrams.

## Gated Dipole Rebounds

In a series of papers beginning in 1972, Grossberg reduced the on-center off-surround architecture of figure 5.1 to the *gated dipole* minimal anatomy. This, in turn, led to a series of remarkable insights into the structure and functioning of mind.

Consider, for example, the rather familiar example at the top of figure 5.2: stare at the black circles for fifteen seconds (longer if the lighting is dim). Then close your eyes. An inverse "retinal afterimage" appears: *white* circles in a *black*

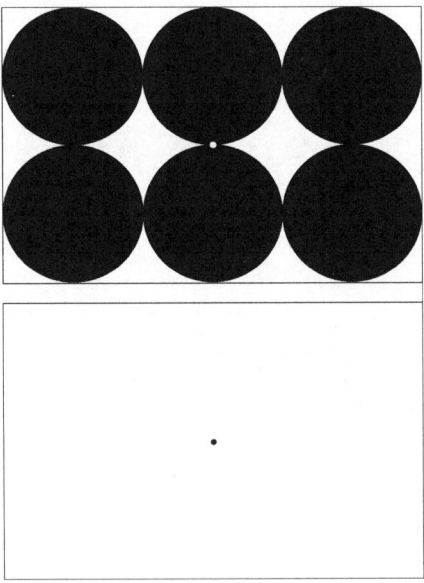

**Figure 5.2.** A McCollough rebound occurs by switching the gaze to the lower pane after habituating to the upper pane.

field![1] Although this percept is often called a "retinal afterimage," it arises mainly in the *lateral geniculate nucleus* of the thalamus and neocortex (Livingstone and Hubel 1987). If, while staring at figure 5.2, a flashbulb suddenly *increases* the illumination, an inverse image *also* appears—and it can occur *during* as well as *after* image presentation. (If you don't have a flashbulb handy, you can simulate this effect by staring at figure 5.2 and then abruptly shifting your gaze to the focusing dot in the center of the all-white at the bottom of figure 5.1.) Both decreasing illumination (closing the eyes) and increasing illumination (the flashbulb effect) can create inverse percepts, and this can happen during, as well as after, a sensation. We can account for all of these effects with the minimal anatomy in figure 5.3.

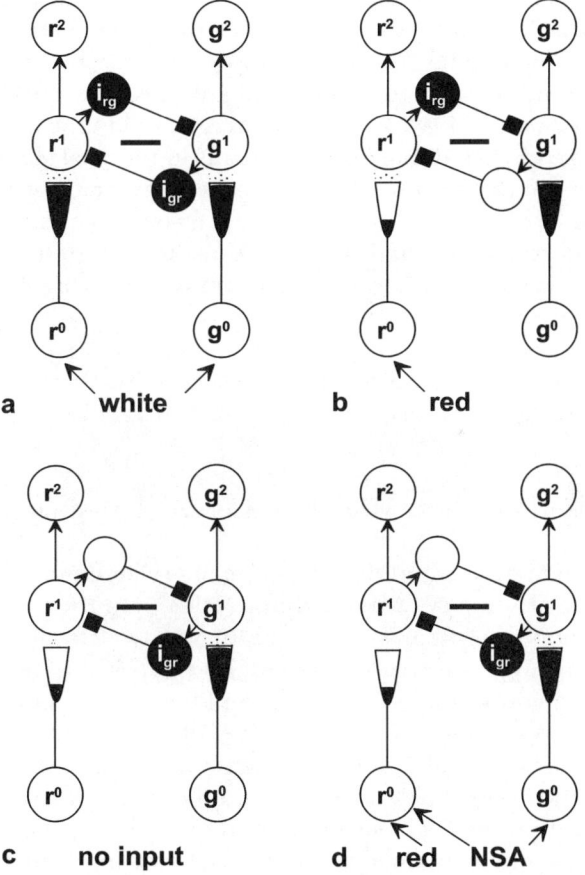

**Figure 5.3.** The McCollough effect. A red-green gated dipole: (a) With white-light input, both poles are active. (b) With red input, the red pole is active and neurotransmitter depletes at the $r^0$–$r^1$ synapse. (c) Closing the eyes allows background activity in the green pole to dominate competition and produce a retinal afterimage. (d) Alternatively, NSA (e.g., a flash of white light) can produce an afterimage rebound, even while red input is maintained.

In the human visual system, black and white, blue and yellow, and red and green response cells are all arrayed in gated dipoles. This leads to a group of phenomena collectively known as the McCollough effect (McCollough 1965; see also Livingstone and Hubel 1987). Under white light, as schematized in figure 5.3a, red and green receptor cells compete to a standoff. White is perceived, but no red or green color percept is independently output. In figure 5.3b, red light illuminates the dipole. The red pole inhibits the green pole via the inhibitory interneuron $i_{rg}$, so only the red pole responds, and a red percept is output from $r^2$. After protracted viewing under intense illumination, however, neurotransmitter becomes depleted at the $r^0$–$r^1$ synapse, and the dipole becomes unbalanced. Neurons maintain a low level of random, background firing even in the absence of specific inputs, so if specific inputs are shut off (e.g., if the eyes are closed), as in figure 5.3c, then the green pole will come to dominate the depleted red pole in response to background activation. On the other hand, in figure 5.3d, a burst of white light stimulates the unbalanced dipole. Because this burst of white light contains equal amounts of red and green light, it is an example of what ART calls *nonspecific arousal* (NSA). Even if the original red input is maintained during this burst, so more red than green remains in the total stimulus spectrum, the green pole still gains control because of its greater neurotransmitter reservoir at synapse $g^0$–$g^1$.[2]

One of Sherrington's many contributions to the understanding of the nervous system was his description of neuromuscular control in terms of *agonist-antagonist* competition. After Sherrington, "antagonistic rebounds," in which an action is reflexively paired with its opposite reaction, began to be found everywhere in neurophysiology. Accordingly, Grossberg referred to events like the red-green reversal of figure 5.3 as "antagonistic dipole rebounds."

## Mathematical Models of Cerebral Mechanics

Grossberg analyzed the gated dipole and many related neural models mathematically, first as a theory of "embedding fields" and then as the more fully developed Adaptive Resonance Theory. Our purpose in the following chapters will be to analyze similar neural models linguistically, but the preceding example offers an opportunity to make a simplified presentation of Grossberg's mathematical models. One of the advantages of mathematical analysis is that, although it is not essential to an understanding of adaptive grammar, it can abstract from the flood of detail we encountered in the previous chapters and bring important principles of system design into prominence. A second reason to develop some mathematical models at this point is that in the second half of this chapter we will use them to build a computer model of a patch of neocortex. We will then be better prepared to explore the question of how language could arise through and from such a patch of neocortex. Finally, many of the leading mathematical ideas of ART are really quite simple, and they can give the nonmathematical reader a helpful entry point into the more mathematical ART literature.

## ART Equations

The central equations of Grossberg's adaptive resonance theory model a *short-term memory trace*, $x$, and a *long-term memory trace*, $z$. Using Grossberg's basic notation, equations 5.1 and 5.2 are differential equations that express the rate of change of short-term memory ($x$ in 5.1) and long-term memory ($z$ in 5.2):

$$\dot{x}_j = -Ax_j + Bx_i z_{ij} \qquad (5.1)$$

$$\dot{z}_{ij} = -Dz_{ij} + Ex_i x_j \qquad (5.2)$$

Equation 5.1 is a differential equation in dot notation. It describes the rate of change of a short-term memory trace, $x_j$. We can think of this short-term memory trace as the percentage of Na$^+$ gates that are open in a neuron's membrane. In the simplest model, $x_j$ decreases at some rate $A$. We can say that $A$ is the rate at which the neuron (or neuron population) $x_j$ "forgets." Equation 5.1 also states that $x_j$ increases at some rate $B$. The quantity $B$ is one determinant of how fast $x_j$ depolarizes, or "activates." We can think of this as the rate at which $x_j$ "learns," but we must remember that here we are talking about *short-term* learning. Perhaps we should say $B$ determines how fast $x_j$ "catches on."

The rate at which $x_j$ catches on also depends upon a second factor, $z_{ij}$, the long-term memory trace. We can think of this long-term memory trace as the size of the synapse from $x_i$ to $x_j$ (cf. the synapses in figure 5.3). Changes in $z_{ij}$ are modeled by equation 5.2. Equation 5.2 says that $z_{ij}$ decreases (or forgets) at some rate $D$, and that $z_{ij}$ also increases at some rate $E$, a function of $x_i$ times $x_j$. We can say that $z_{ij}$ "learns" (slowly) at the rate $E$.

It is important to note that $A$, $B$, $D$, and $E$ are all shorthand abbreviations for what, in vivo, are very complex and detailed functions. The rate $B$, for example, lumps all NMDA and non-NMDA receptor dynamics, all glutamate, aspartate, and GABA neurotransmitters, all retrograde neurotransmitters and messengers, all neurotransmitter release, reuptake, and manufacture processes, membrane spiking thresholds, and who knows what else into a single abstract function relating barrel $x_j$ to barrel $x_i$ across synapse $z_{ij}$. This may make $B$ seem crude (and undoubtedly it is), but it is the correct level of abstraction from which to proceed.

It is also important to note that, by self-similarity, ART intends $x_j$ and $z_{ij}$ to be interpretable on many levels. Thus, when speaking of an entire gyrus, $x_j$ might correlate with the activation that is displayed in a brain scan. When speaking of a single neuron, $x_j$ can be interpreted as a measure of the neuron's activation level above or below its spiking threshold. When speaking of signal propagation in the dendritic arborization of a receptor neuron, $x_j$ can be intereprented as the membrane polarization of a dendritic branch. In these last two cases, a mathematical treatment might explicitly separate a threshold function $\Gamma$ out from equation 5.1, changing $+Bx_i z_{ij}$ into something like $+B\Gamma(x_i, z_{ij})$. Usually, however, ART equations like 5.1 and 5.2 describe neural events on the scale of the barrel or of larger, self-similar subnetworks. At these scales, $x_i$ may have dozens or thousands of pyramidal output neurons, and at any particular moment, 3 or 5 or 5000 of them may be above spiking threshold. The subnetwork as a whole will have a *quenching threshold,* above which the activity of neurons in the subnetwork will be amplified and below which the activity of neurons in the subnetwork will be attenuated. But the subnetwork as a whole need not exhibit the kind of "all-or-none" discrete spiking threshold that has been claimed for individual neurons, so ART equations do not usually elaborate a term for thresholds. Instead, they use nonlinear gating functions.

## Nonlinearity

There is only one way to draw a straight line, but there are many ways to be nonlinear, and nonlinearity holds different significance for different sciences. ART equations are nonlinear in two ways that are especially important to cognitive modeling: they are (1) sigmoidal and (2) resonant.

The curves described by ART equations and subfunctions (like $A$, $B$, $D$, and $E$ above) are always presumed to be *sigmoidal* (∫-shaped). That is to say, they are naturally bounded. For example, a neuron membrane can only be activated until every $Na^+$ channel is open; it cannot become more activated. At the lower bound, a membrane can be inhibited only until every $Na^+$ channel is closed; it cannot become more inhibited.[3] So $x_j$ has upper and lower limits, and a graph of its response function is sigmoidal. In equation 5.2, $z_{ij}$ is similarly bounded by the sigmoidal functions $D$ and $E$.

Equations 5.1 and 5.2 also form a nonlinear, *resonant* system. The LTM trace $z_{ij}$ influences $x_j$, and $x_j$ influences $z_{ij}$. Both feedforward and feedback circuits exist almost everywhere in natural neural systems, and feedforward and feedback circuits are implicit almost everywhere in ART systems: for every equation 5.1 describing $x_j$'s response to $x_i$, there is a complementary equation 5.1' describing $x_i$'s reciprocal, *resonant,* response to $x_j$.[4] This is the same kind of nonlinearity by which feedback causes a public address system to screech out of control, but equations 5.1 and 5.2 are bounded, and in the neural systems they describe, this kind of feedback makes rapid learning possible.

## Shunting

The fact that the terms of ART equations are multiplicative is an important detail which was not always appreciated in earlier neural network models. Imagine that table 5.1 presents the results of four very simple paired associate learning experiments in which we try to teach a parrot to say $ij$ (as in "h-i-j-k"). In experiment A, we teach the parrot $i$ and then we teach it $j$. The parrot learns $ij$. In experiment B, we teach the parrot neither $i$ nor $j$. Not surprisingly, the parrot does not learn $ij$. In experiment C, we teach the parrot $i$, but we do not teach it $j$. Again, it does not learn $ij$. In experiment D, we do not teach the parrot $i$, but we do teach it $j$. Once again, it does not learn $ij$.

TABLE 5.1. Truth table for logical AND (multiplication).

|   | A | B | C | D |
|---|---|---|---|---|
| $i$ | 1 | 0 | 1 | 0 |
| $j$ | 1 | 0 | 0 | 1 |
| Learned? | 1 | 0 | 0 | 0 |

The not-very-surprising results of our parrot experiment clearly reflect the truth table for *multiplication*. So ART computes learning by multiplying $x_i$ by $x_j$ in equation 5.2. Similarly, in equation 5.1, ART multiplies $x_i$ by $z_{ij}$. In the literature on artificial neural networks and elsewhere, engineers often refer to such multiplicative equations as *shunting* equations.

## Habituation

In the psychological literature *habituation* is said to occur when a stimulus ceases to elicit its initial response. The rebound described in figure 5.3 is a common example of a habituation effect. Although the term is widely and imprecisely used, we will say that habituation occurs whenever neurotransmitter becomes depleted in a behavioral subcircuit. Neurotransmitter depletion can be described with a new equation:

$$\dot{n}_{ij} = + Kz_{ij} - Fn_{ij}x_i \tag{5.3}$$

Equation 5.3 states that the amount of neurotransmitter $n$ at synapse $ij$ grows at some rate $K$, a function of $z_{ij}$, the capacity of the long-term memory (LTM) trace. By equation 5.3, $n_{ij}$ is also depleted at a rate $F$, proportional to the presynaptic stimulation from $x_i$. Put differently, $z_{ij}$ represents the *potential* LTM trace, and $n_{ij}$ represents the *actual* LTM trace. Put concretely, at the scale of the single synapse, $z_{ij}$ can be taken to represent (among other factors) the available NMDA receptors on the postsynaptic membrane, while $n_{ij}$ represents (among other factors) the amount of presynaptic neurotransmitter available to activate those receptors. Given equation 5.3, equation 5.1 can now be elaborated as 5.4:

$$\dot{x}_j = -Ax_j + B\sum_{i \neq j} n_{ij}x_i - C\sum_{k \neq j} n_{kj}x_k \tag{5.4}$$

Equation 5.4 substitutes actual neurotransmitter, $n_{ij}$, into the original $Bx_iz_{ij}$ term of equation 5.1. It then elaborates this term into two terms, one summing multiple excitatory inputs, $+B\Sigma n_{ij}x_i$, and the second summing multiple inhibitory inputs, $-C\Sigma n_{kj}x_k$. This makes explicit the division between the long-distance, on-center excitatory inputs and the local, off-surround inhibitory inputs diagrammed in figure 5.1.

Habituation, specific and nonspecific arousal, and lateral inhibition, as described in equations 5.2–5.4, give rise to a computer model of cerebral cortex and a range of further cognitive phenomena, including noise suppression, contrast enhancement, edge detection, normalization, rebounds, long-term memory invariance, opportunistic learning, limbic parameters of cognition, P300 and N400 evoked potentials, and sequential parallel memory searches. These, in turn, form the cognitive basis of language.

## A Quantitative Model of a Cerebral Gyrus

Figure 5.4a–e shows what happens when equations 5.2–5.4 are used to create a computer model of a cerebral gyrus.[5] The model gyrus in figure 5.4 is twenty-three barrels high by forty-eight barrels wide. Each barrel forms forty-eight excitatory synapses with other barrels at radius 3–4 and twenty-four inhibitory synapses with twenty-four barrels at radius 1–2. The gyrus is modeled as a closed system in the shape of a torus: barrels at the top edge synapse with the bottom edge, and barrels at the left edge synapse with the right edge. Each synapse's $z_{ij}$ and $n_{ij}$ are modeled by equations 5.2 and 5.3, while each barrel's $x_j$ is modeled by equation 5.4. Figure 5.4 displays the activation level of each barrel (its $x_j$ value) according to the gray scale at the bottom: black is least active and white is most active.

At time $t = 0$ in figure 5.4a, the gyrus is an inactive, deep-gray tabula rasa. Specific inputs $I$ are applied to target nodes at [$x$ $y$] coordinates [10 9], [10 11], [10 13], and [10 15], and a black, inhibited surround begins to form. At $t = 1$ after another application of specific inputs to the target field, resonant activation begins to appear at radius 3–4 (figure 5.4b).

### Noise suppression

At time $t = 1$ (figure 5.4b), the target nodes are activated above the level of the rest of the gyrus, and a black, inhibitory surround has formed around them. The inhibitory surround is graphic illustration of how *noise suppression* arises as an inherent property of an on-center off-surround system: any noise in the surround that might corrupt the on-center signal is suppressed.

### Contrast enhancement

Figure 5.4b also illustrates *contrast enhancement*, a phenomenon similar to noise suppression. The target nodes at $t = 1$ are light gray; that is, their activity is contrastively "enhanced" above the background and the target nodes at $t = 0$ (figure 5.4a). This enhancement is not only due to repeated, additive specific inputs; it is also due to resonant feedback *to* the target node field *from* the emerging active fields at radius 3–4 and the lateral inhibition just described under noise suppression.

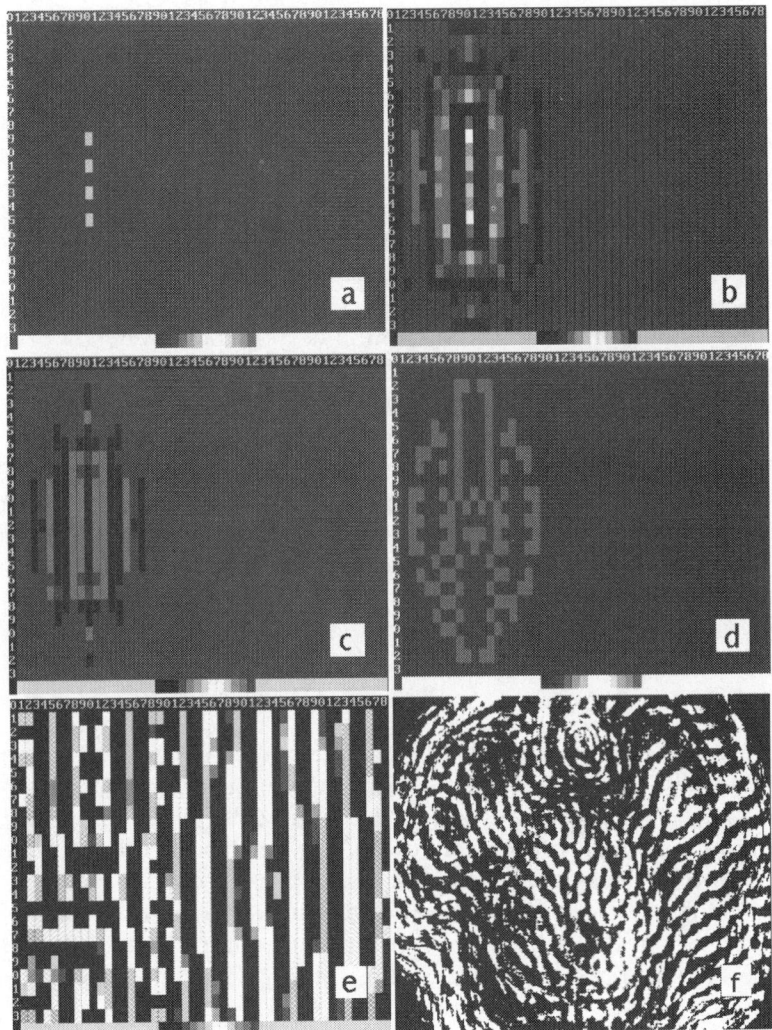

**Figure 5.4.** (a–b) A cerebral gyrus simulated by equations 5.2–5.4. (c–e) A cerebral gyrus simulated by equations 5.2–5.4. (f) A gyrus of macaque striate cortex radiographically labeled by tritium. (LeVay et al. 1985. Reprinted by permission of the Society for Neuroscience.)

Edge detection

In figure 5.4b, inputs continue to be applied through $t = 1$, and a resonant pattern begins to develop. *Edge detection* emerges as another inherent property of the ART system: the non-edge nodes at [10 11] and [10 13] are actively inhibited on both flanks (by each other, as well as by the edge nodes [10 9] and [10 15]). On the other hand, the edge nodes are each actively inhibited only by nodes on one interior flank (by [10 11] or [10 13]). Consequently, the edges are less inhibited, more active, and more perceptually detectable.

Normalization

A fourth property of on-center off-surround anatomies is *normalization*. Consider as an example the intensity of speech sounds. For human hearing, the threshold of pain is somewhere around 130 dB. What happens at a rock concert, where every sound may be between 120 dB and 130 dB? In the on-center off-surround anatomy of auditory cortex, if one is not first rendered deaf, a frequency component at 130 dB will inhibit its neighbors. A neighboring frequency component at 120 dB can thus be inhibited and perceived as if it were only 80 or 90 dB. In this way, an on-center off-surround anatomy accomplishes a kind of automatic gain control—normalization that prevents the system from saturating at high input levels.

On-center off-surround anatomies accomplish a similar kind of normalization in vision. At sunset, daylight is redder than it is at noon. Nevertheless, at sunset we still perceive white as white, not as pink. In this case, an on-center off-surround perceptual anatomy keeps the world looking normal even as its optical characteristics physically change.

Rebounds

Figure 5.4c displays NSA applied at $t = 2$ to all nodes in the gyrus. This is analogous to the "flashbulb effect" discussed with reference to figure 5.3. After NSA, at $t = 3$ (figure 5.4d), a *rebound* occurs: the target cells, which previously were "on" relative to their surround, are now "off." Note that the rebound does not only occur on the local scale of the original target barrels. A larger scale, self-similar rebound can also occur over the left and right hemifields of the gyrus (fields [$x$ 1–20] vs. [$x$ 21–48]). Rebounds occur on large, multicellular scales just as they occur on smaller, cellular scales. This capacity for fieldwide rebounds is critical to preserving long-term memory invariance.

Complementation and long-term memory invariance

Long-term memory is, by definition, invariant: A pattern once learned (like that in figure 5.4a) should not be easily forgotten. But amid all the busy, buzzing, resonant neural activity suggested by figure 5.4, what is to prevent a remembered pattern like that of figure 5.4a from being overwritten by other, conflicting inputs? Rebounds are the answer to this cognitive problem, too. Suppose that I ask you to learn the pattern in figure 5.4a, but after you have studied it for a little while, I change my mind and say, "No, stop. Now learn this second pattern." In later chapters we will see how my *No* instruction causes NSA and a general rebound, putting your mind in the tabula-not-quite-rasa state of figure 5.4d. If, while you are in this new state, I present you with a new pattern to learn, the original target nodes will be inactive. By equation 5.2, synaptic learning at those nodes will also be inactivated. (Let the rebounded, inhibited target nodes, $x_r$, in figure 5.4e have short-term memory (STM) activation levels of 0, i.e., $x_r = 0$. Then by equation 5.2, $Ex_i x_r = 0$, so all $z_{ir}$ (0.) In this way, rebounds *complement* memory, partitioning it and preventing new information from overwriting old, long-term memories.

Ocular dominance columns

If the resonance begun in figure 5.4a–b is allowed to resonate without conflicting inputs, the gyrus eventually achieves the pattern of figure 5.4e. Figure 5.4e bears a striking resemblance to *ocular dominance columns* in primary visual (striate) cortex (figure 5.4f).

Wiesel and Hubel (1965; Wiesel et al. 19974) sutured one eye of newborn kittens closed. After several months, this caused neocortical cells that were wired to the sutured eye to become permanently unresponsive. Similar suturing of adult cats had no such effect, so Wiesel and Hubel proposed that there existed a *critical period* during which vision had to become established by experience. Subsequent radiographic staining techniques allowed Wiesel, Hubel, and Lam (1974) and others to make dramatic pictures showing that eye-specific striate cortex cells were arranged in stripes, or "columns." Figure 5.4f is one such picture, in which white stripes are neurons responding to one eye (not sutured but radiographically labeled by tritium; LeVay et al. 1985). Like other sensory systems, the visual system had been known to exhibit a retinotopic mapping from peripheral receptors in the eye up through cortex. Ocular dominance columns appeared as one more remarkable instance of such topographical mapping. (The *tonotopic* mapping of the auditory system will figure prominently in our next several chapters on speech and speech perception.)

Hubel and Wiesel's studies exerted a broad influence in science, and Lenneberg (1967) proposed that a similar "critical-period" explanation could be extended to language. Coupled with Chomsky's speculations on the innateness of language, Lenneberg's thesis suggested that there existed a detailed genetic plan for grammar, just as there could be supposed to exist a detailed genetic plan in which segregation of thalamocortical axons formed the anatomic basis for detailed ocular dominance columns. Because only about $10^5$ human genes are available to code the $10^8$-odd axons innervating striate cortex (not to mention Broca's area, Wernicke's area, and all the rest of the human brain), this suggestion was never adequate, but in the absence of better explanations it was widely accepted.

We will return to these issues of critical periods and neuronal development in later chapters. For now, it remains only to establish that the similarities between our model gyrus and real neocortex are not fortuitous. To this end, note that the similarities between figure 5.4e and figure 5.4f are not only impressionistic and qualitative but also quantitative: the diameter of the stripes in figure 5.4e is approximately two barrels—the radial extent of a barrel's inhibitory surround in the model gyrus. It also happens that the width of Wiesel and

Hubel's ocular dominance columns is 0.4 mm—approximately the radial extent of cerebral inhibitory cells' inhibitory surround.[6]

## XOR

The potential advantages of massively parallel computers have been vaguely apparent for quite some time. In fact, the first large-scale parallel computer, the 64,000-element Illiac IV, became operational in the mid-1960s. But before the invention of microchips, parallel machines were prohibitively expensive, and once built, the Illiac IV proved extremely difficult to program. The leading parallel-computing idea of the day was Rosenblatt's *perceptron* model (Rosenblatt 1958, 1959, 1961), but in 1969 Minsky and Papert's widely influential book *Perceptrons* (1969; see also Minsky and Papert 1967) claimed to prove that perceptrons were incapable of computing an XOR (see below). As a result, they argued, perceptrons were incapable of calculating *parity*, and therefore incapable of performing useful computational tasks.[7] Because of this argument, XOR has figured prominently in subsequent parallel-computing theory. We will return to this issue, but for the present, consider only that *dipoles calculate XOR*.

XOR, or "exclusive OR," means "A or B but not both." Formally, it is a Boolean logical operation defined for values of 1 and 0 (or true and false). For the four possible pairs of 1 and 0, XOR results in the values given in table 5.2. XOR is 1 (true) if A or B—but not both—is 1. This is the same function that gated dipoles computed in figure 5.3 and 5.4. Grossberg (1972a, passim) found that gated dipoles compute XOR ubiquitously in the brain, as an essential and natural function of agonist-antagonist relations and lateral inhibition. Calculating parity is no longer thought essential to computation, but as we have seen and as we shall see, dipoles and XOR are essential to noise suppression, contrast enhancement, edge detection, normalization, long-term memory invariance, and a host of other indispensable properties of cognition.

### Opportunistic Learning with Rebounds

The rate of neurotransmitter release from the presynaptic terminal is not constant. When a volley of signal spikes releases neurotransmitter, it is initially released at a higher rate. With repeated firing, the rate of release of neurotransmitter decreases. As suggested previously, this constitutes synaptic habituation (see also Klein and Kandel 1978, 1980).

TABLE 5.2. Truth table for exclusive OR (XOR).

| | | | | |
|---|---|---|---|---|
| A: | 1 | 1 | 0 | 0 |
| B: | 1 | 0 | 1 | 0 |
| XOR | 0 | 1 | 1 | 0 |

A plausible mechanical explanation for this habituation is that while a knob is inactive, neurotransmitter accumulates along the synaptic membrane (see figure 3.7). When the first bursts of a signal volley depolarize this membrane, the accumulated transmitter is released. The synapse receives a large burst of neurotransmitter and a momentary advantage in dipole competition. Thereafter, transmitter continues to be released, but at a steadily decreasing rate. Meanwhile, the inhibited pole of a dipole is dominated, but it is not slavish. All the while it is dominated, it accumulates neurotransmitter reserves along its synaptic membranes, preparing itself for an opportunistic rebound.

Rebounds make for opportunistic learning in a manner reminiscent of Piagetian *accommodation* (Piaget 1975). Learning implies the learning of *new* information, so if old information is not to be overwritten when the new information is encoded, a rebound complements and repartitions memory to accommodate the new information.

## Expectancies and Limbic Rebounds

A few pages back it was suggested that a rebound could be caused by a teacher saying, "No, wait. Learn a different pattern." But what if there is no teacher? For billions of years, intelligence had to survive and evolve in the "school of hard knocks"—without a teacher. How could rebounds occur if there were no teacher to cause them?

In figure 5.4b, we saw secondary, resonant fields form, encoding the input pattern presented in figure 5.4a. In a richer environment, these secondary resonant fields encode not only the immediate input but also, simultaneously, the associated context in which input occurs. In the future, the learned $z_{ij}$ traces of context alone can be sufficient to evoke the target pattern. These secondary fields are *contextual expectancies,* or, as Peirce put it in 1877, "nervous associations—for example that habit of the nerves in consequence of which the smell of a peach will make the mouth water" (9).

In 1904, Pavlov won the Nobel Prize for discovering (among other things) that associations could be conditioned to arbitrary contexts. For example, if one rings a bell while a dog smells a peach, one can later cause the dog's mouth to water by just ringing the bell. In 1932, Tolman gave such abstract contexts the name we use here—*expectancies*. Grossberg's suggestion (1980, 1982a) was that failed expectancies could trigger rebounds, thereby making it possible for animals and humanoids to learn new things during the billion years of evolution that preceded the appearance of the first teacher.

## Evoked Potentials

As we have seen, the action potentials of neurons are electrical events, and in 1944, Joseph Erlanger and Herbert Gasser received the Nobel Prize for developing electronic methods of recording nerve activity. One outgrowth of their work was the ability to measure *evoked potentials*—the voltage changes that a stimulus evokes from a nerve or group of nerves. Allowing for variation among different nerve groups, a fairly consistent pattern of voltage changes has been observed in response to stimuli. During presentation of a stimulus sentence like

$$\text{Mary had a little lamb, its fleece was white as coal} \qquad (5.5)$$

$$\text{CNV} \qquad\qquad\qquad\qquad \text{P100 P300 N400}$$

a series of voltage troughs and peaks labeled CNV, P100, P300, and N400 may be recorded by scalp electrodes positioned over language cortex. CNV, or *contingent negative variation* is a negative voltage trough that can be measured as a usual correlate of a subject's expectation of a stimulus, for example, sentence 5.5. A positive peak, P100, can then be recorded 100 ms after stimulus onset, when a stimulus is first perceived. If, however, the stimulus is unrecognized—that is, if it is *unexpected*—a P300 peak is reliably evoked, approximately 300 ms after stimulus onset. The rapid succession of syllables in a sentence like 5.5 can obscure these potentials on all but the last word, but in cases of anomalous sentences like 5.5 (Starbuck 1993), there is a widespread N400 "anomaly component" which is clearly detectable over associative cortex.

Grossberg's theory (1972b, 1980; Grossberg and Mcrrill 1992) is that P300 corresponds to a burst of NSA that is triggered by the collapse of an expectation resonance. More specifically, Grossberg (1982a) suggests that old knowledge and stable expectancies have deep, subcortical resonances extending even to the hippocampus and limbic system. If events in the world are going according to the mind's plan, then heart rate, breathing, digestion, and all the other emotions and drives of the subcortical nervous system are in resonance with the perceived world, but a disconfirming event, whether it be the unexpected attack of a predator, a teacher's *No!* or the simple failure of a significant expectancy causes this harmonious resonance to collapse. This collapse unleashes (disinhibits) a wave of NSA, causing a rebound and making it possible for the cerebrum to accommodate new information.

Grossberg's P300 theory is also compatible with Gray's "comparator" theory of hippocampal function, and together they give a satisfying explanation of certain facts discussed in chapter 4, like HM's anterograde amnesia. According to this Gray-Grossberg theory, HM could not learn new things because, with a resected hippocampus, he could not compare cerebral experience against his subcerebral emotional state and so could not generate rebounds. Without rebounds, his cerebral cortex could not be partitioned into active and inactive sites, so inactive sites could not be found to store new memories.

Although HM could not remember new names and faces, he could retain certain new, less primal long-term memories such as the solution to the Tower of Hanoi puzzle. By the Gray-Grossberg theory, HM might have been able to retain these memories because nonemotional experiences are relatively nonlimbic and may therefore depend to a lesser extent on limbically generated NSA. As we saw from figure 5.3, NSA is not the only way to trigger a rebound.

Turning down inputs by shifting attention (or closing the eyes) can generate a rebound just like turning up inputs. This is why we are often able to solve a problem by "sleeping on it." Turning off the inputs that have been driving our cognitive processes allows alternative hypotheses to rebound into activity and successfully compete for LTM storage.

Grossberg's P300 hypothesis suggests that Gray's comparator theory might be tested by evoked potentials, but unfortunately, the hippocampus and other key limbic centers lay deep within the brain, inaccessible to measurement by scalp electrodes and even by sophisticated devices like *magnetoencephalograms* (MEG scans). On the other hand, measurement techniques which can "go deep," like *positron emission tomography* (PET) scans and *functional magnetic resonance imaging* (fMRI) scans, resolve events on a scale of seconds at best—far off the centisecond timescale on which evoked potentials occur.

Sequential parallel search

Repeated rebounds can enable a sequential parallel search of memory. If a feedforward input pattern across some field $F^1$ does not match (that is, resonate with) an expected feedback pattern across $F^2$, a burst of NSA can cause a rebound across $F^2$, and a new feedback pattern can begin to resonate across $F^2$. If this new $F^2$ expectancy does not match the input, yet another rebound can cause yet a third feedback pattern to be instantiated across $F^2$. This process can repeat until a match is found.

Peirce (1877) described this process as *abduction*, which he contrasted with the classical logical processes of induction and deduction. He exemplified it with the story of Kepler trying to square Copernicus's theoretical circular orbits with the planets' occasional retrograde motion. Kepler tried dozens of hypotheses until he finally hit upon the hypothesis of elliptical orbits. While Kepler's trial-and-error abductive process was not logical in any classical sense, it did reflect the logic of expectancies, trying first the expected classical, Euclidean variants of circular orbits before finding resonance in the theory of elliptical orbits.

Such searches are not random, serial searches through all the $10^{7,111,111}$-odd patterns in the mind. Rebounds selectively occur in circuits which have minimal LTM (minimal neurotransmitter reserves) and maximal STM activity. By weighing experience against present contextual relevance, the search space is ordered, and—unlike a serial computer search through a static data structure—abduction quickly converges to a best resonance.

In this chapter we have modeled cerebral dynamics using only a handful of basic differential equations. These equations nevertheless enabled us to build a computer simulation which exhibits subtle cognitive dynamics and surprising similarities to real mammalian neocortex. Many of these dynamic properties are especially evident in language phenomena, to which we will turn in chapter 7. First, however, chapter 6 must give the reader a quick survey of the physics and physiology of speech and hearing.

• S I X •

# Speech and Hearing

In the previous chapters we reviewed the central nervous system and introduced adaptive resonance theory (ART) as a system for modeling it. Most of the data presented in those chapters dealt with vision. Language, however, deals primarily with sound. Therefore, in this chapter we will first consider some essential, physical facts of speech. We will then see how these physical data are processed by the ear and the auditory nervous system into the signals that the cerebrum ultimately processes as language.

## Speech

### Periodic sounds

Speech may be divided into two types of sounds: periodic and aperiodic. These types correspond roughly to the linguistic categories of vowel and consonant. We will focus on these categories because our first linguistic neural networks (in chapter 7) will model how we perceive them.

Vowels are the prototypic periodic speech sounds. They originate when air is forced from the lungs through the glottis, the opening between the vocal cords of the voice box, or larynx (figure 6.1).[1] When air flows between them, the vocal cords vibrate. In figure 6.2, a single string, which could be a violin string, a piano string, or a vocal cord, is shown vibrating.

Each time the string in figure 6.2 stretches to its right, air molecules are set in motion to the right. They bump against adjacent molecules and then bounce back to the left. In this manner a chain reaction of compressed "high-pressure areas" is set in motion to the right at the speed of sound. (Note that molecule $A$ does not wind up at position $Z$ 1 second later. Rather, one must imagine that molecule $B$ bumps molecule $C$ to the right. Molecule $B$ then bounces back and bumps molecule $A$ to the left. Eventually, $Y$ bumps $Z$. When

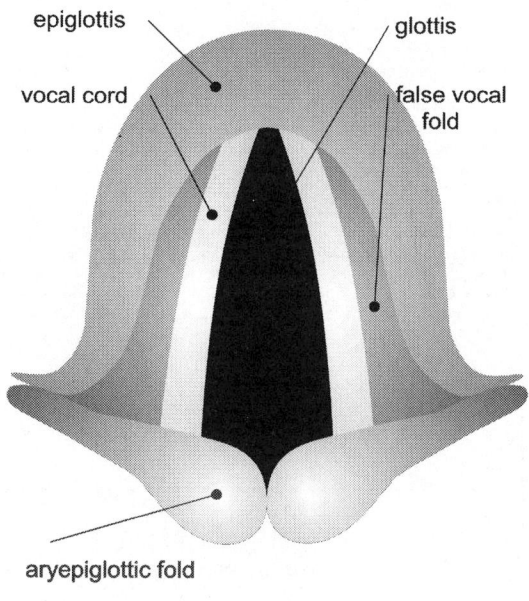

**Figure 6.1.** The larynx.

the string springs back to the left, alternating low-pressure areas are formed between the high-pressure areas. The resulting sound waveform spreads like the waves from a pebble dropped in a still pond. If we plot these high- and low-pressure areas (figure 6.2), we get a sinusoidal (sine-wave-shaped) sound wave. These waves are commonly seen when a microphone transduces high and low air pressure waves into electrical signals displayed on an oscilloscope.

The ordinate of the plot (AMP in figure 6.2) shows how the pressure rises and falls. This measure of rise and fall is the wave's *amplitude*. It corresponds to the sound's intensity or loudness. (The human ear does not hear very low or very high sounds, so technically, "loudness" is how intense a sound *seems*.) In figure 6.2 the wave is shown to complete two high-low cycles per second. The wave's measured frequency in figure 6.2 is therefore two *hertz* (2 Hz).[2]

Every string has a natural fundamental frequency ($f_0$), which depends upon its length and elasticity. When a string of given length and elasticity vibrates as a whole, as in figure 6.2, it produces the *first harmonic* ($H_1$). But if we divide the string in half by anchoring its midpoint, as in figure 6.3, each half-string will vibrate twice as fast as the whole. The plot of the resulting high- and low-pressure areas will be a wave with twice the frequency of $f_0$. Figure 6.3 is drawn to show such a wave at 4 Hz, twice the fundamental frequency in figure 6.2. In musical terms, $H_2$ sounds an octave higher than the fundamental frequency, $f_0$. Note, however, that each half-string in figure 6.3 moves less air than the whole string of figure 6.2, so the amplitude of $H_2$ is less than that of $H_1$.

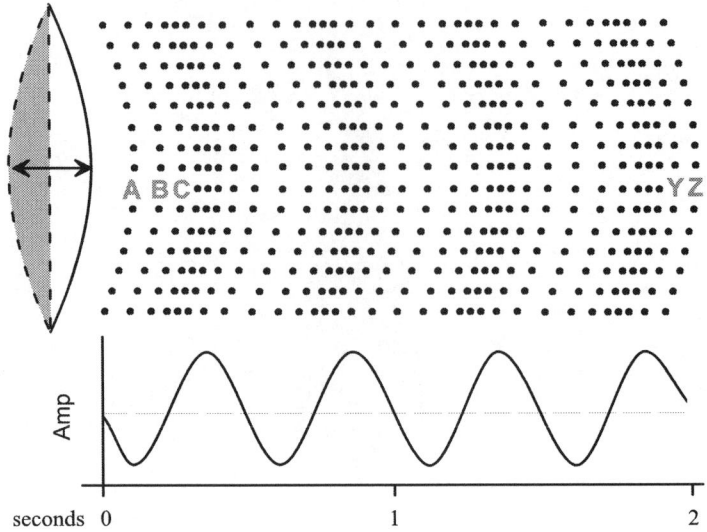

**Figure 6.2.** The first harmonic of a vibrating string and its waveform.

Now things become complex. It happens that a string can vibrate in many segments at once. In figure 6.4, our string is shown vibrating as a whole to the right. At the same time, each third is also vibrating, the top and bottom thirds to the right, and the middle third to the left. Each third vibrates at three times the frequency of the whole. A string's vibration can be (and usually is) more complex still. A classic example occurs if a complex wave is composed of suc-

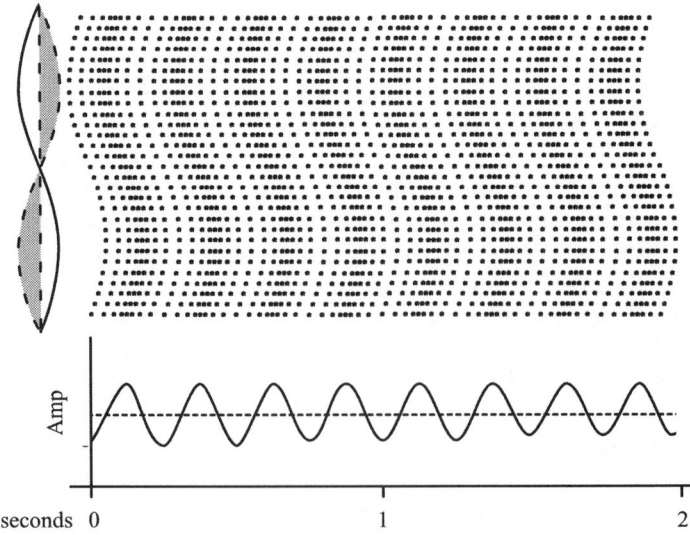

**Figure 6.3.** A string vibrating in halves generates the second harmonic.

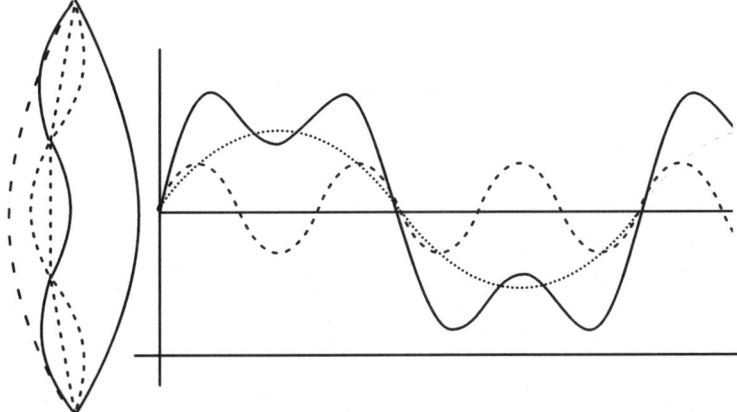

**Figure 6.4.** $H_1$ (dotted line) and $H_2$ (dashed line) in complex vibration.

cessive odd harmonics (e.g., the first, third, and fifth). In this case, the composite complex wave approaches a square wave in shape.

When two or more harmonics are present, they can reinforce each other or interfere with each other. When they reinforce each other, their peaks align and become higher; the harmonics are said to *resonate* and the sound becomes louder. For example, when the dotted-line and dashed-line harmonics in figure 6.4 combine, the first peak of the resulting solid-line waveform is more intense than either of the two other components alone. In highly resonant systems, a waveform can actually *feed back* into itself, resonating with itself and becoming louder and louder. This is what happens, for example, when a public address system microphone is placed too close to its speakers. The output feeds back into the microphone, reinforcing and resonating with itself until a fuse blows. When sound waves interfere with one another, their energy dissipates and the vibration of the string is described as *damped:* once plucked, it does not continue to vibrate.

This is the situation with the vocal cords. Their vibratory patterns are very complex and not very resonant. The resulting sound wave—the *glottal pulse*—is a highly damped waveform (figure 6.5). A single glottal pulse sounds much more like a quick, dull thud than a ringing bell.

Each time a puff of air passes from the lungs through the larynx, one glottal pulse like figure 6.5 occurs. The vocal folds then spring closed again until subglottal air pressure builds to emit another puff. If, while you were talking, we could somehow unscrew your head just above your larynx, your larynx would sound like a "Bronx cheer" (a bilabial trill). This is approximately the sound of a trumpet mouthpiece without the trumpet. What turns such a buzzing sound into melodious, voiced speech?

A trumpet makes its buzzing mouthpiece harmonious by attaching it to a tube. If we put a pulse train like figure 6.5 into a tube, a chain reaction of high- and low-pressure areas is set up in the tube. Like strings, tubes also have harmonic frequencies at which they resonate. Like strings, the shorter the tube,

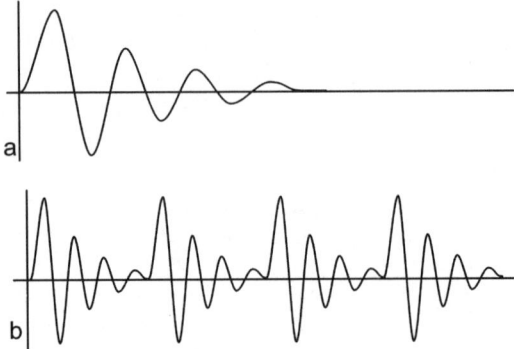

**Figure 6.5.** (a) The glottal pulse has a highly damped waveform. (b) Continuous periodic speech is based on a train of glottal pulses.

the higher the fundamental frequency. If we now screwed your head back onto your larynx, your normal, beautiful, melodious voice would be restored. Your mouth and throat function like the tubes of a trumpet to transform a lowly Bronx cheer into vowels. To gain a deeper appreciation of how this happens, we need to visualize sound waves in the frequency domain.

In figures 6.2–6.4, we observed that a string could simultaneously vibrate at successively higher harmonic frequencies, each of successively lower amplitude. Figure 6.6 captures these facts in a display known as a *power spectrum*. Each vertical line in figure 6.6a represents a harmonic in figure 6.4, and the height of each line represents the amplitude of the harmonic. The amplitudes decrease as the harmonics increase in frequency. Notice that figure 6.6a represents essentially the same physical facts as figure 6.4: a complex wave with $f_0 = 1$ Hz. and $H_1 = 3$ Hz.

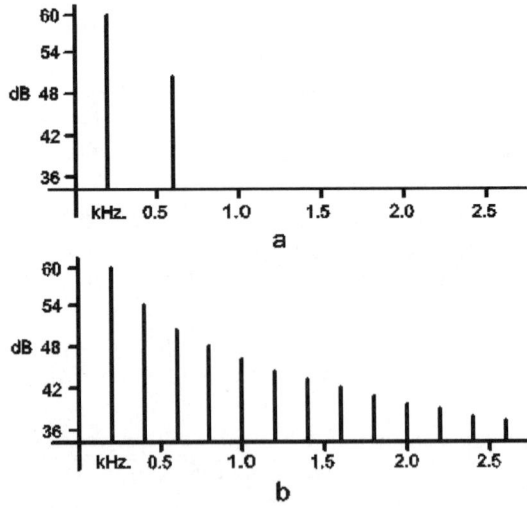

**Figure 6.6.** Power spectrum of (a) figure 6.4 and (b) the glottal pulse.

Figure 6.4, however, plots time on the *x*-axis, and is therefore sometimes called a *time domain* plot. Figure 6.6 plots frequency on the *x*-axis, and so is called a *frequency domain* plot. Figure 6.6b presents a fuller picture of the frequency domain of the glottal pulse, whose many harmonics fall off in amplitude as frequency increases.

Figure 6.7 explains the operation of a trumpet starting with a frequency domain plot of the buzzing input at the mouthpiece (the vertical lines). Let the fundamental frequency of figure 6.7 be a train of pulses at 200 Hz. Then higher harmonics are generated at integer multiples of this $f_0$, that is, at 400 Hz, 600 Hz, 800 Hz, and so on. The tube of the trumpet, however, imposes a further *filter* that "envelopes" this input. With some specific combination of valves, the tube of the trumpet takes on a length which resonates optimally at some note at the center frequency of the bell curve in figure 6.7. The surrounding frequencies resonate less and are attenuated. Most of them are completely filtered out and never escape the trumpet. On the other hand, the tube can actually amplify the resonant frequencies at the center of the curve. When the trumpeter changes the valves on a trumpet, tubes of different lengths are coupled together. These successive tubes of differing length are resonant frequency filters which select the successive musical notes the trumpeter is playing.

The human voice works rather like a trumpet trio. The glottal pulse train feeds into three connected tubes of variable length. These resonate at three different, variable frequencies and produce the various three-note, chordlike sounds, which are called vowels. These three "tubes" can be seen in figure 6.8, which shows the articulatory position of the vocal tract for the vowel /i/, as in *bead*.

Letters enclosed in brackets and slashes, like [i] and /e/, are letters of the International Phonetic Alphabet, or IPA, which is useful for distinguishing sounds like the vowels of *bad* and *bade*. IPA departs from normal English spelling in several respects, most notably in that /i/ represents the sound of French *si* or Chinese *bi*, but English *deed*, and /e/ represents the sound of French *les* or German *Schnee*, but English *bade*. Letters enclosed in brackets are *phones*, and letters enclosed in slashes are *phonemes*. Phones are sounds-in-the-air, and phonemes are sounds-in-the-mind.

The phoneme /i/ is often described as a high, front vowel, meaning the tongue is high and to the front. This position of the tongue divides the vocal

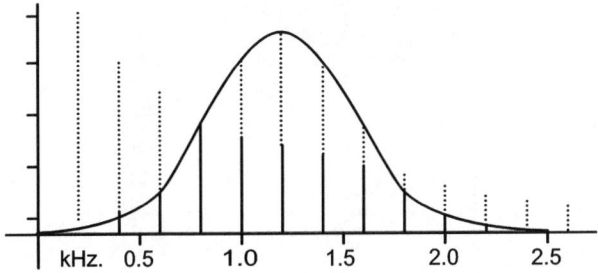

**Figure 6.7.** Idealized trumpet filter centered at 1.2 kHz.

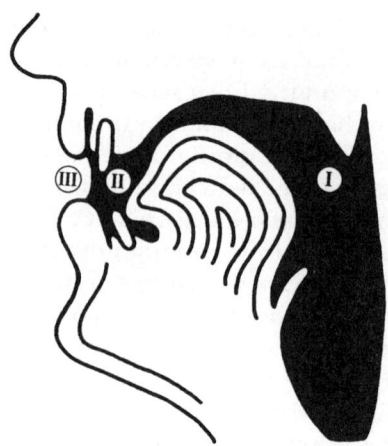

**Figure 6.8.** Articulatory position of /i/.

tract into three subtubes (figure 6.8): the labiodental cavity, between the teeth and the lips (III); the oral cavity, in the middle (II); and the pharyngeal cavity at the rear (I). The dental cavity resonates at a frequency that defines the highest note, or *formant* ($F_3$), of the vocalic "chord."[3] The oral cavity is larger. When articulating /i/, the oral cavity is bounded at the rear by the tongue and at the front by the teeth. Together with the labiodental cavity, the oral cavity forms a longer, larger subtube which resonates at a lower frequency, the second formant, $F_2$, of the vowel. The $F_1$ formant is the lowest in frequency. It is created by the pharyngeal, oral, and labiodental cavities resonating together as one large, long tube.

For comparison, figure 6.9 shows the articulatory position for the vowel /u/, as in *rude*. Here, the tongue is up and back, making the oral cavity bigger. The lips are also rounded, making the labiodental cavity bigger. Correspond-

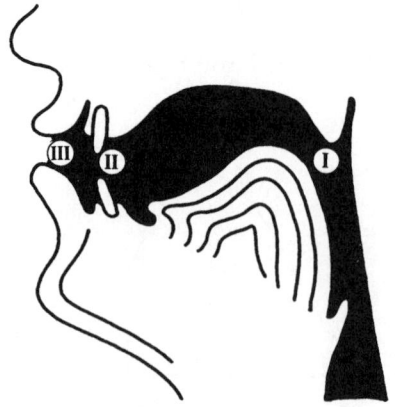

**Figure 6.9.** Articulatory position for /u/.

ingly, $F_2$ and $F_3$ are both lower than in /i/. $F_1$ is low as in /i/ because the overall length and volume of the vocal tract remain large.

Figure 6.10 summarizes the main acoustic effects of the articulatory configurations in figures 6.8 and 6.9, superimposing filter functions (or *vowel spectra*) on a male glottal spectrum. In figure 6.10a, the spectrum for the vowel /i/ displays typical formant peaks at 300, 2200, and 2900 Hz. For this speaker, these are the three "notes" of the vocalic "chord" for /i/, and they correspond to the resonant frequencies of the three subtubes of the vocal tract in figure 6.8. Corresponding typical formant values for /u/ of 300, 750, and 2300 Hz are plotted in figure 6.10b, values corresponding to the three subtubes of the vocal tract in figure 6.9.

The preceding three-tube model is adequate for our purposes, but it is also a considerable oversimplification. The acoustics of coupled tubes is actually much more complex. Retroflex consonants like American /r/ divide the oral cavity into two resonant chambers, and lateral semivowels like /l/ define the oral cavity with two apertures, one on each side of the tongue. Nasal speech sounds reso-

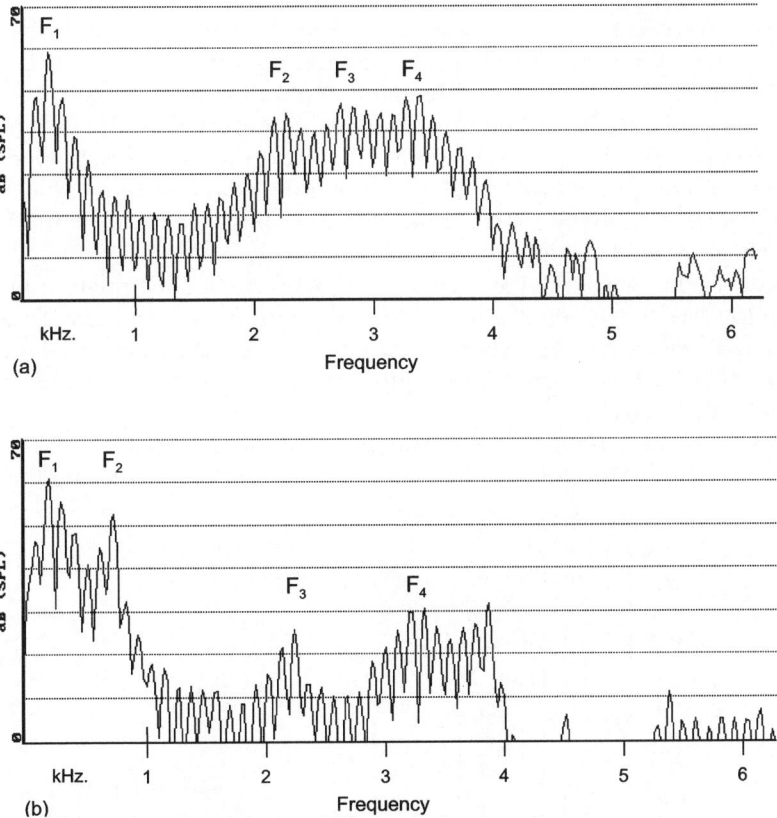

**Figure 6.10.** Glottal spectra and vocal tract filters for (a) /i/ and (b) /u/ (fast Fourier transform analyses).

nate in complex nasal and sinus cavities and introduce antiresonances which cannot be accounted for in our simple model. Because everyone's vocal tract is different, there is also endless variation among individual voices.

One interesting source of individual variation is illustrated in figure 6.11. Males have lower voices than women and children, so a male with a fundamental frequency of 100 Hz will have glottal source harmonics at 200, 300, 400 Hz, and so on (all vertical lines). However, a female with a fundamental frequency of 200 Hz will have only half as many harmonics, at 400, 600, 800 Hz, and so on (dotted and solid vertical lines, dotted formant envelope). Since the vocal tract can only amplify harmonics that are physically present in the source spectrum, the formants of female speech are less well defined than the formants of male speech. This becomes especially apparent if the female fundamental frequency in figure 6.11 is raised an octave to 400 Hz. Now only the solid-line harmonics are present in the source. There is no harmonic at 1400 Hz, and the second formant, which the vocal tract locates at that frequency, barely resonates (solid-line formant envelope). This makes one wonder why mommies, rather than daddies, teach language to children. We will explain later how mommies compensate for this apparent disadvantage.

A child's vocal tract is shorter still, so its formants are higher, and this compensates somewhat for the child's higher $f_0$. But as a result of its shorter vocal tract, all the child's vowel formants fall at frequencies different from mommy's formants (not to mention daddy's and everybody else's). This "lack of invariance" has been an enormous frustration to attempts at speech recognition by computer, and it makes one wonder that children can learn language at all. Later we will also see how the child's brain normalizes this variance to make language learning possible.

A device known as the "sound spectrograph" can make time-frequency domain plots like figures 6.12 and 6.13 directly from speech and expose some of this variation. In a standard spectrogram, the *x*-axis is the time axis. Frequency is plotted on the *y*-axis. Amplitude is plotted "coming at you" on the *z*-axis, with louder sounds darker.

Figure 6.12a is a narrowband spectrogram of the vowel /i/ (as in *deed*). The harmonics of the glottal source appear as narrow bands running the length

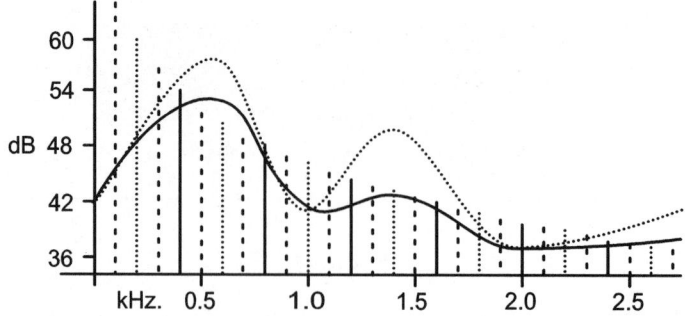

**Figure 6.11.** A high voice decreases formant definition.

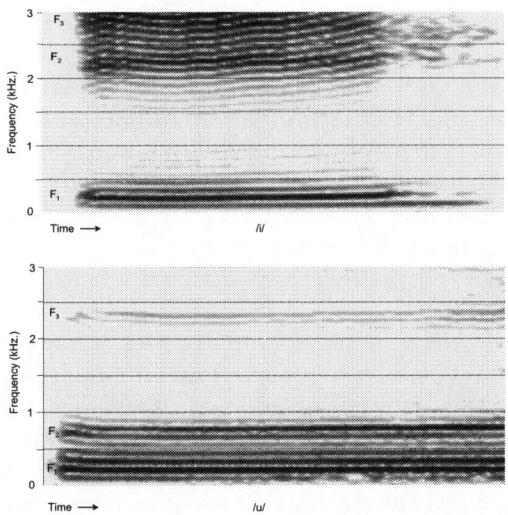

**Figure 6.12.** Narrowband spectrograms of (a) /i/ and (b) /u/.

of the spectrogram. The three characteristic formants of /i/ appear as dark bands of harmonics which are resonantly amplified by the vocal tract filter. (Sometimes higher fourth and fifth formants appear, as in figure 6.10, but they carry little speech information.) Figure 6.12b is a spectrogram of the vowel /u/ (as in *crude*). The phoneme /u/ differs from /i/ in that the second formant is much lower, at 800 Hz.

Aperiodic sounds

Aperiodic sounds are sounds that do not have a fundamental frequency or harmonic structure. For example, figure 6.13 shows a spectrogram of the aperiodic sound /s/, as in *sassy*. All aperiodic sounds in speech can be said to be consonants, but not all consonants are aperiodic. For example, nasal sounds (e.g., English /m/, /n/, /ŋ/) and certain "semivowels" (e.g., English /l/, /r/,

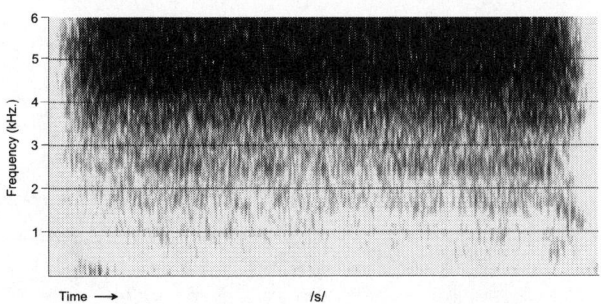

**Figure 6.13.** Spectrogram of /s/.

/w/) are actually periodic, and the argument can be made that such "consonants" should properly be called vowels. In any event, it follows that "consonant" is something of a catchall term, and the features of consonants cannot be summarized as succinctly as we have summarized the articulatory and acoustic structure of vowels. Unlike vowels, however, many consonants are quick sounds, which make them especially interesting examples of the motor control of speech.

Motor control of speech

Figure 6.15 identifies key articulatory points in the linguistic dimension of *place*. For example, /b/ and /p/ are articulated at the lips; their place is said to be *labial*. English /d/ and /t/ are said to be *alveolar*; they are articulated by placing the tongue tip at the *alveolus*, the gum ridge, behind the upper incisors. By contrast, /g/ is *velar*. It is produced by placing the back of the tongue against the *velum*, the soft palate at the back of the roof of the mouth. Acoustically, these consonants are identified by *formant transitions*, changes in the following (or preceding) vowel's formants, which disclose how the articulators have moved and the vocal tract has changed to produce the consonant.[4] Figure 6.14 illustrates place acoustically with spectrograms of [bɑb], [dɑd], and [gɑg].

Initially, [b], [d], and [g] are all *plosive* consonants, which begin with a vertical line representing a damped and brief *plosive burst*. This is most clearly visible for [d(d] in figure 6.14. Bursts are damped and brief because they have competing, not-very-harmonic "harmonics" at many frequencies. These bursts are followed by the formant transitions, which may last only 50 ms. The transitions appear as often-faint slopes leading into or out of the steady-state vowel formants. They are traced in white in figure 6.14. The transitions begin (for initial consonants) or end (for postvocalic consonants) at frequencies which roughly correspond to the consonant's place of articulation (Halle et al. 1957).

For example, [b] is a *labial* consonant, meaning it is articulated at the lips. Since the lips are closed when articulation of [bɑb] begins, high frequencies are especially muted and all formant transitions appear to begin from the fun-

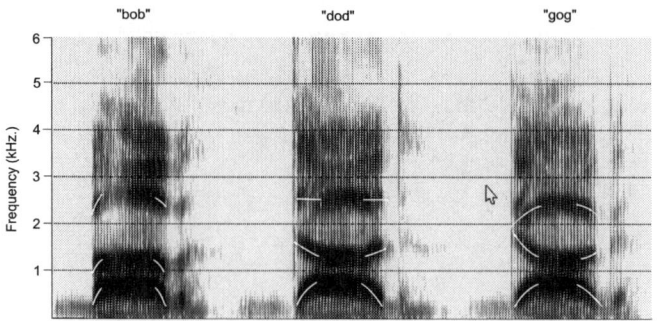

**Figure 6.14.** Spectrograms of [bɑb], [dɑd], [gɑg].

damental frequency (figure 6.14). The transitions of [d] begin at frequencies near those for the vowel [i] (figure 6.14), because [i] is a "high, front" vowel, with the tongue tip near the alveolar ridge, and [d] is articulated with the tip of the tongue at the alveolar ridge. Accordingly, [d] is given the place feature *alveolar*. Similarly, [g] begins its formant transitions near the frequencies for the vowel [u] (figure 6.14). The vowel [u] is produced with the body of the tongue raised back toward the velum, and [g] begins its articulation with the tongue raised back in contact with the velum. Therefore, [g] is classed as a *velar* consonant.[5]

The different ways sounds can be articulated at any given place is described by the articulatory feature of *manner*. Thus, before vowels, [d] and [g] are called *plosive* consonants. They display their common manner of articulation in figure 6.14 by beginning with an explosive burst. But after vowels, their manner of articulation is somewhat different and they are called *stops*. (In older literature the distinction between stops and plosives was not appreciated, and both were often simply called stops.)

The *fricative* consonant [s] results from an intermediate gesture. The tongue is placed too close to the alveolus to produce [i] but not close enough for a stop. Instead, the tongue produces friction against the outgoing airstream. This friction causes a turbulent airflow and the aperiodic sound [s], which is classed as +*fricative*. In spectrograms, [s] can exhibit formant transitions from and to its surrounding vowels, but its aperiodic sound and fricative manner produce a more distinctive and salient band of wideband "noise" (figure 6.13). Fricative sounds articulated further forward in the mouth with a smaller frontal cavity, like [f] and [θ] (as in *thin*), generate a higher frequency noise spectrum. Sounds generated further back in the mouth with a larger frontal cavity, like the alveopalatal [ʃ] in *shoot*, generate a lower frequency noise spectrum.

As the preceding examples should illustrate, very delicate movements of tongue, lips, jaw, velum, and glottis must be coordinated to produce any single speech sound. In fluent speech, these movements must also be quick enough to produce as many as twenty distinct phones per second, and at such high rates of speed, articulation can become imprecise. Fortunately, the vocal tract has optimal locations, certain places of articulation, like the alveolus and velum (figure 6.15), that are highly fault-tolerant. At these locations, articulation can be maximally sloppy, yet minimally distort the resulting speech sound. It is not surprising that sounds produced at these "quantal" locations, including the "cardinal vowels" [i], [a], [u], occur in virtually all human languages (Stevens 1972).

This quantal advantage is leveraged by the upright posture of *Homo sapiens* (Lieberman 1968). Animals that walk on all fours, apes, and even slouched Neanderthals all have short, relatively straight vocal tracts. When humans started walking erect, the vocal tract lengthened and bent below the velum, creating separate pharyngeal and oral cavities. This added an additional formant to human calls, essentially doubling the number of speech sounds humans could produce. The already highly innervated and developed nervous system for eating was then recruited to help control this expanded inventory, and human speech evolved.

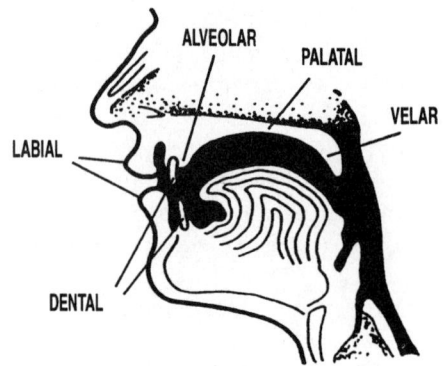

**Figure 6.15.** Optimal places of articulation.

## Hearing

The ear

The human ear (figure 6.16) can be divided into three parts: the outer ear, including the ear canal, or *meatus*; the middle ear; and the inner ear. The middle ear is composed of three small bones: the "hammer" (malleus), the "anvil" (incus), and the "stirrup" (stapes). The inner ear is composed of the vestibular system, which senses up and down, and the cochlea, the primary organ for hearing.

As we saw in figure 6.10, vowel formants are mostly between 300 and 3000 Hz. Most information is conveyed around 1800 Hz, a typical mean frequency for $F_2$. Having described the speech system as a kind of trumpet, we can now see the outer ear to be a kind of ear trumpet (figure 6.16). This outer-ear assembly has evolved so that it has a tuning curve (figure 6.17) centered on a resonant frequency of 1800 Hz. Natural selection has tuned the outer ear to speech (or at least to the range of human "calls": ape ears are very similar even though we might not wish to say that apes "speak"). One reason the ear must

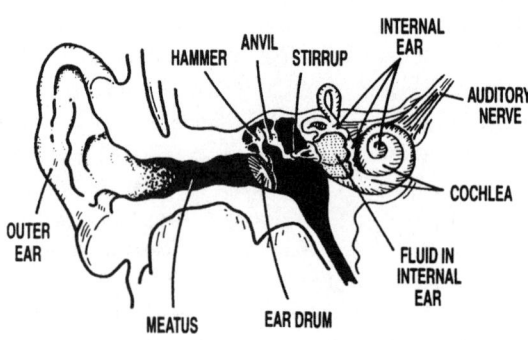

**Figure 6.16.** The ear.

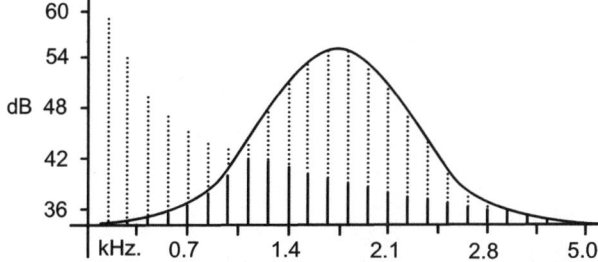

**Figure 6.17.** The tuning curve of the outer ear.

be carefully tuned to speech is that sound waves are feeble. The sound pressure of whispered speech might be only 0.002 dynes/cm². The ear must transform this weak physical signal into something palpable and sensible. What is equally remarkable is that the ear can also respond to sound pressure levels up to 140 dB, a 10,000,000-fold increase over the minimum audible sound.

Sound waves cause the eardrum to vibrate. In turn, the three small bones of the middle ear act as levers which amplify and transmit these vibrations to the oval window of the cochlea. More important, the oval window is much smaller than the eardrum, further amplifying the pressure delivered to the cochlea. With this middle-ear assembly, we are able to hear sounds 1000 times fainter than we could otherwise. (It also contains small muscles which protect us from sounds that are too loud.)

The oval window is a membrane "opening into" the cochlea. The cochlea is an elongated cone, coiled up like a snail shell (and so called *cochlea*, from the Greek *kochlias*, "snail"). In 1961, Georg von Békésy received the Nobel Prize for describing how these vibrations from the eardrum are ultimately sensed by the cochlea (Békésy 1960). The cochlea is fluid-filled, and the vibrations the middle ear transmits to this fluid are transferred to the basilar membrane, which runs inside the cochlea from the oval window to its tip. As the basilar membrane vibrates, it produces a shearing action against *hair cells*, which are the end-organ cells of the auditory nervous system. Each hair cell emits a nerve signal (a spike) when its hair is stimulated by the shearing action of the basilar membrane.

Figure 6.18 schematically uncoils the cochlea and exposes the basilar membrane. We can envision the hair cells arrayed along this membrane like the keys of a mirror-image piano, high notes to the left. (Unlike a piano, the broad end of whose soundboard resonates to low frequencies, the membrane's thin, flexible tip resonates to low frequencies, while the more rigid broad end resonates to high frequencies.) When the spectrum of a vowel like [a] stimulates the basilar membrane, it sounds like a chord played on the hair cells. Signals from the activated hair cells then propagate along the *auditory nerve*: from the cochlea along the cochlear nerve to the cochlear nucleus, the inferior colliculus, the medial geniculate body of the thalamus, and the transverse temporal gyrus of the cerebrum (koniocortex), as illustrated in figure 6.19.

**Figure 6.18.** The cochlea as a piano soundboard.

The cochlear nucleus

The cochlear nucleus is not *in* the cochlea. It is the first auditory processing substation along the auditory nerve. In the cochlear nucleus, different cell types and networks process the auditory signal in at least four distinctive ways (Harrison and Howe 1974; Kelly 1985).

First, large spherical neurons in the cochlear nucleus not only relay the hair cells' signals *but do so tonotopically.* That is, they preserve the cochlea's regular, keyboardlike mapping of different frequencies. Each spherical cell responds to only one narrow frequency range because each is innervated by only a few hair cells. Like narrowband spectrograms, these neurons transmit periodic, vocalic information to higher processing centers.

Second, *octopus cells* in the cochlear nucleus (figure 6.20) respond to signals from wide frequency ranges of hair cells. Because they sample a broad

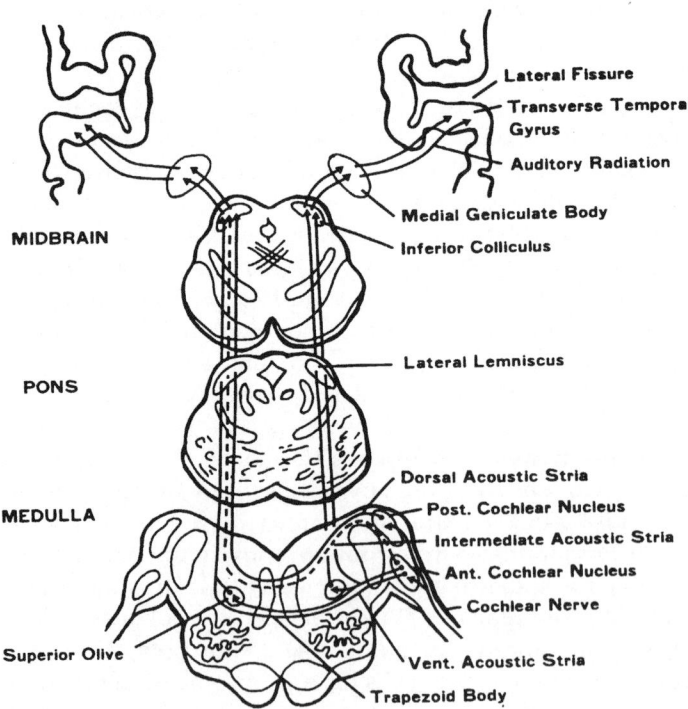

**Figure 6.19.** Central auditory pathways. (Manter 1975. Reprinted by permission of F. A. Davis Company.)

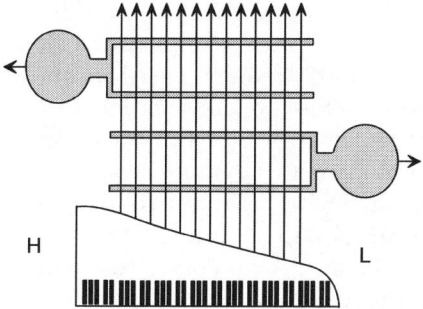

Figure 6.20. Octopus cells of the cochlear nucleus.

spectrum, octopus cells are well designed to function like sound meters, measuring the overall intensity of a complex sound.

Third, because octopus cells receive inputs from many hair cells at many frequencies at once, they are capable of quickly depolarizing in response to brief stimuli like the many frequencies in a plosive burst. By contrast, it takes a relatively long time for the few hair cells in a single narrow frequency band to depolarize a spherical cell.

Fourth, many octopus cells' dendrites are arrayed to receive inputs either from high to low or from low to high (figure 6.20). Thus, these cells are differentially sensitive to brief events such as the rising and falling formant transitions which mark place of articulation (figure 6.14; Kelly 1985).

From the cochlear nucleus afferent auditory axons enter the *trapezoid body* and cross over to the *superior olivary nucleus* (superior olive) on the contralateral side of the brain stem. However, unlike vision and touch, not all auditory pathways cross. An *ipsilateral* (same-side) pathway also arises (i.e., from left ear to left cerebral hemisphere and from right ear to right cerebral hemisphere). Thus, each side of the brain can compare inputs from both ears. Sounds coming from the left or right side reach the left and right ears at slightly different times, allowing the brain to identify the direction of sounds. Bernard Kripkee has pointed out to me how remarkable the ability to localize sound is. We learned in chapter 3 that brain cells fire at a maximum rate of about 400 spikes per second. Nevertheless, the human ear can readily distinguish sound sources separated by only a few degrees of arc. At the speed of sound, this translates into time differences on the order of 0.0004 s. This means the auditory system as a whole can respond about 100 times faster than any single neuron in it. Instead of the Von Neumannesque, all-or-nothing, digital response of a single neuron, many neurons working together in parallel produce a nearly analogue response with a hundredfold improvement in sensory resolution.

Ultimately, pathways from both the cochlear nucleus and the superior olive combine to form the *lateral lemniscus*, an axon bundle which ascends to the *inferior colliculus*. The inferior colliculus (1) relays signals to higher brain centers (especially the *medial geniculate nucleus of the thalamus*, MGN) and (2) in doing so preserves the tonotopic organization of the cochlea. That is, the topographic arrangement by frequency which is found in the cochlea is repeated in the inferior colliculus. (And indeed is repeated all the way up to the cerebrum!) The inferior colliculus has been clearly implicated in sound localization, but little is known about the functions of the inferior colliculus with respect to speech. However, it is noteworthy that reciprocal connections exist both from the medial geniculate nucleus (MGN) and from the cerebral cortex back to the inferior colliculus, and these pathways will prove important to one of our first adaptive grammar models in chapter 7.

Like the inferior colliculus, the medial geniculate nucleus relays afferent auditory signals to auditory cortex, retaining tonotopic organization. It also receives reciprocal signals from auditory cortex and sends reciprocal signals back to the inferior colliculus. Unlike the inferior colliculus, the medial geniculate nucleus also exchanges information with thalamic centers for other senses, especially vision. As a result, some cross-modal information arises from thalamus to cortex. Moreover, much of this information seems to be processed in cerebrum-like on-center off-surround anatomies.

Medial geniculate nucleus projections of the auditory nerve erupt into the cerebrum in the (auditory) koniocortex on the inner surface of the superior temporal gyrus, inside the Sylvian fissure. Perhaps because this area is relatively inaccessible to preoperative brain probes, it has been relatively little studied in the human case. Nevertheless, it can be inferred from dissection, as well as from many studies of mammals and primates, that koniocortex is characterized by tonotopic neuron arrays (*tonotopic maps*) which still reflect the tonotopic organization first created by the cochlea. Considerable research on tonotopic mapping has been done on mammalian brains ranging from bats (Suga 1990) to monkeys (Rauschecker et al. 1995). In fact, these brains tend to exhibit three, four, five, and more cerebral tonotopic maps.

Since the same on-center off-surround architecture characterizes both visual and auditory cortex, the same minimal visual anatomies from which adaptive resonance theory was derived can be used to explain sound perception. We will first consider how auditory contrast enhancement may be said to occur, and then we will consider auditory noise suppression.

Auditory contrast enhancement

At night, a dripping faucet starts as a nearly inaudible sound and little by little builds until it sounds like Niagara Falls, thundering out all possibility of sleep. Figure 6.21 models this common phenomenon. Field $F^1$ (remember that fields are superscripted while formants are subscripted) models a tonotopic, cochlear nucleus array in which the drip is perceived as a single, quiet note activating cell $x$ subliminally ("below the limen," the threshold of perception). For concreteness, the higher field $F^2$ may be associated with the inferior colliculus or the medial geniculate nucleus. The response at $F^1$ also graphs the (subliminal) response at $t_1$, while the response at $F^2$ also graphs the (supraliminal) response at some later $t_2$. At $t_1$ the response at cell $x$, stimulated only by the faucet drip, barely rises above the surrounding stillness of the night and does not cross the

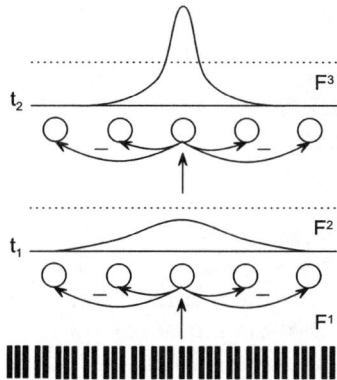

**Figure 6.21.** Auditory contrast enhancement.

threshold of audibility. However, cell $x$ is stimulated both by the drip and by resonant stimulation. At the same time, the surrounding cells, ..., $x-2$, $x-1$ and $x+1$, $x+2$, ..., become inhibited by cell $x$, and as they become inhibited, they also *disinhibit* cell $x$, adding further to the on-center excitation of $x$. This process ("the rich get richer and the poor get poorer") continues until finally, at $t_n$, cell $x$ stands out as loudly as any sound can. The contrast between the drip and the nighttime silence has become enhanced.

Auditory noise suppression and edge detection

Auditory noise suppression is closely related to contrast enhancement since both are caused by the dynamics of on-center off-surround neural anatomies. Noise suppression and an interesting, related case of spurious edge detection are illustrated in figure 6.22.

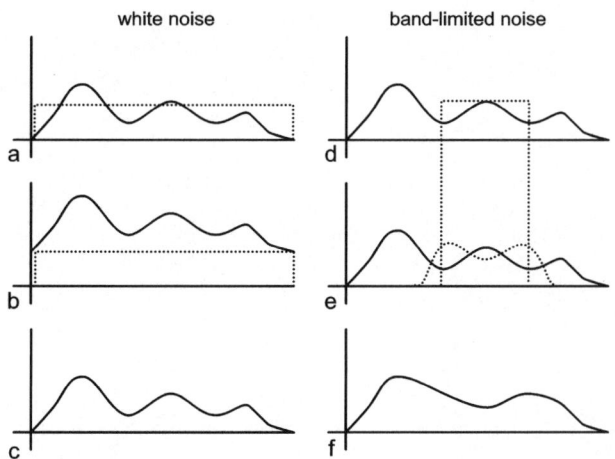

**Figure 6.22.** White noise is suppressed; band-limited noise is not.

In World War II communications research, it was found that white noise interfered with spoken communication less than band-limited noise. Figure 6.22 presents a vowel spectrum (solid line) under conditions of both white and band-limited noise (dotted lines). At the same amplitude, when noise is limited to a relatively narrow frequency band, there is "less" of it than there is of white noise, which covers the entire frequency range. Nevertheless, the band-limited noise interferes more with speech communication. This occurs because, under white noise, the on-center off-surround neural filter causes the noise to be uniformly suppressed across the entire spectrum. At the same time, contrast enhancement picks out the formant peaks and emphasizes them above the background noise. Thus, in figure 6.22, under the white-noise condition, the perceived formant spectrum is preserved at time $c$, after neural processing. The band-limited noise, however, introduces perceptual edges which, like the edges in figure 5.4b, become enhanced. By time $f$ (figure 6.22), the speech spectrum has been grossly distorted: the perceptually critical second formant peak has been completely suppressed and replaced by two spurious formant peaks introduced at the edges of the band-limited noise. This further illustrates the perceptual phenomenon of *edge detection*, which occurs because in on-center off-surround processing, the middle of the band-limited noise is laterally inhibited and suppressed from both sides, while the edges are suppressed only from one side.

In this chapter, we have looked at the basic human motor-speech and auditory systems, and we have seen how these systems produce, sample, sense, and process sound and perceive it as the most fundamental elements of speech. In particular, we examined basic examples of noise suppression, contrast enhancement, and edge detection in on-center off-surround ART networks. In chapter 7 we will extend these findings from these atomic, *phonetic* speech sounds and phenomena to the *phonemic* categories of speech.

• S E V E N •

# Speech Perception

In chapter 6 we described the nature of the speech signal and how its image is sensed and presented to the cerebrum. We noted that because no two vocal tracts are exactly alike, your pronunciation will differ subtly but certainly from my pronunciation. To express these subtle, *phonetic* differences, linguists invented the *International Phonetic Alphabet* (IPA). In the fifteenth century, the Great English Vowel Shift caused the writing system of English to deviate from continental European systems, so IPA looks more like French or Italian spelling than English. Thus, when you say *beet*, we might write it in IPA as [bit], and if I pronounce my [i] a little further forward in my mouth, we could capture this detail of my pronunciation in IPA as [bi⁺t]. The sounds of the letters of IPA correspond to *phones*, and the brackets tell us we are attempting to capture pronunciation *phonetically*, that is, as accurately as possible.

But even though you say [bit] and I say [bi⁺t], we both perceive the word *beet*. This is a small miracle, but it is very significant. In one sense, language itself is nothing but a million such small miracles strung end to end. No two oak trees are exactly alike either, but 99% of the time you and I would agree on what is an oak and what isn't an oak. This ability to suppress irrelevant detail and place the objects and events of life into the categories we call words is near to the essence of cognition. In linguistics, categorically perceived sounds are called *phonemes*, and phonemic categories are distinguished from phonetic instances by enclosing phonemes in slashes. Thus, for example, you say [bit] and I say [bi⁺t], but we both perceive /bit/.

Whereas in chapter 6 we found much to marvel at in the neural *production* of phones, we turn now to the still more marvelous phenomena of categorical neural *perception* of phonemes. To understand speech perception, and ultimately cognition, we now begin to study how minimal cerebral anatomies process the speech signal. The result of this exercise will be the first collection of hypotheses that define adaptive grammar.

## Voice Onset Time Perception

The words *beet* and *heat* differ by only their initial phonemes: /bit/ versus /hit/. Linguists call such pairs of words *minimal pairs*. More minimally still, *beet* and *peat* only differ on a single feature. The phonemes /b/ and /p/ are both bilabial plosive (or stop) consonants. The only difference is that /b/ is a *voiced* consonant, and /p/ is an *unvoiced* consonant. The words /bit/ and /pit/ differ only on the manner feature of *voicing*. Such minimal pairs isolate the categorical building blocks of language and provide an excellent laboratory in which to begin the study of cognition.

The difference between voiced and unvoiced sounds was long said to be that the vocal cords vibrated during production of voiced sounds but not during the production of unvoiced sounds. This is true enough, but as we have seen, the *production* of speech sounds is only one-half of the language equation. Following the invention of the sound spectrograph in the late 1940s it became possible for researchers to study the other half, the *perception* of speech sounds. In a landmark study, Liberman et al. (1952) used spectrography to measure the plosive voicing contrast against how it is *perceived* by listeners. For example, spectrograms of /p/ and /b/ in *paid* and *bade* are given figure 7.1.

The spectrograms for both /p/ and /b/ begin with a dark, vertical band which marks the initial, plosive burst of these consonants. (We will examine the third spectrogram in figure 7.1 a little later in this chapter.) These are followed by the dark, horizontal bands of the formants of /e/. Finally, each spectrogram ends with another burst marking the final /d/. It is difficult to find much to say about nothingness, so one might believe, as linguists long did, that the most significant difference between /p/ and /b/ is the aspiration following /p/. This is the high-frequency sound in figure 7.1, appearing after the burst in *paid*. From a listener's perspective, however, such aspiration falls outside the tuning curve of the ear canal and is too faint to be reliably heard. It might as well be silence. And indeed, in 1957 Liberman et al. found that it was the *silence* following a plosive burst which distinguished /p/ and /b/. They called this silent interval *voice onset time* (VOT).

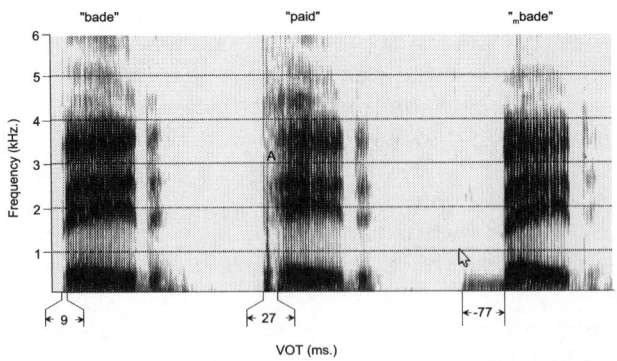

**Figure 7.1.** Spectrograms of [bed], [ped], and [ₘbed] (Spanish-like prevoicing).

Marked VOT in figure 7.1, voice onset time is usually measured from the burst to the beginning of the lowest dark band on the spectrogram, the *voicing bar*. Once researchers recognized that silence could be a highly contrastive speech cue, it was easy to see from spectrograms how VOT could be a highly salient feature of speech. Using synthesized speech, numerous studies quickly verified that VOT was the primary feature distinguishing voiced and unvoiced consonants and that this distinction applied universally across languages (Lisker and Abramson 1964).

It was soon discovered that these perceptual distinctions were also *categorical*. In 1957, Liberman et al. presented listeners with a series of syllables which varied in VOT between 0 and 50 ms (in IPA we might represent the stimuli as [ba], [b$^+$a], [b$^{++}$a], etc.). They asked listeners to identify these syllables as either /ba/ or /pa/. As figure 7.2 shows, they found an abrupt, categorical shift toward the identification of /pa/ when VOT reached 25 ms. Initial plosives with VOTs under 25 ms were perceived as voiced, while those with longer VOTs were perceived as unvoiced. It was as if a binary switch flipped when VOT crossed the 25 ms boundary.

This metaphor of a binary switch was particularly attractive to generative philosophers, who viewed language as the product of a computational mind, and the metaphor took on the further appearance of reality when Eimas, et al. (1971) demonstrated that even extraordinarily young infants perceived the voiced-voiceless distinction categorically. In a series of ingenious studies, Eimas and his coworkers repeated synthetic VOT stimuli to infants as young as one month.

In these experiments, the infants were set to sucking on an electonically monitored pacifier (figure 7.3). At first, the synthetic speech sound [ba]$_0$ (i.e., VOT = 0) would startle the neonates, and they would begin sucking at an elevated rate. The [ba]$_0$ was then repeated, synchronized with the infant's sucking rate, until a stable, baseline rate was reached: the babies became habituated to (or bored with) [ba]$_0$.

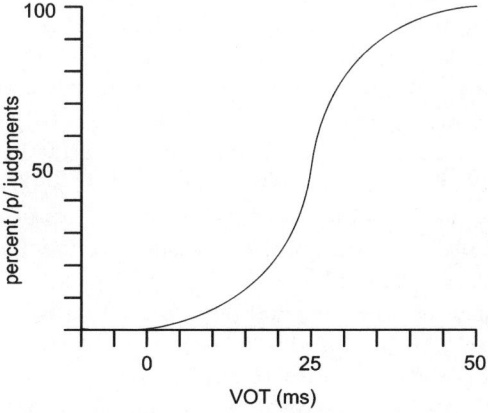

**Figure 7.2.** Categorical perception.

112 • HOW THE BRAIN EVOLVED LANGUAGE

**Figure 7.3.** Eimas et al.'s (1971) "conjugate sucking" paradigm.

Then the stimulus was changed. If it was changed to $[ba]_{30}$, the infants were startled again. They perceived something new and different, and their sucking rate increased. If, however, the new stimulus was $[ba]_{10}$ or $[ba]_{20}$ and did not cross the 25 ms. VOT boundary, the babies remained bored. They perceived nothing new, and they continued to suck at their baseline rate.

This study was replicated many times, and the conclusion seemed inescapable: Chomsky's conjecture on the innateness of language had been experimentally proved. Neonates had the innate capacity to distinguish so subtle and language-specific a feature as phonemic voicing! But then the study was replicated once too often. In 1975, Kuhl and Miller replicated the Eimas study—but with chinchillas! Obviously, categorical perception of VOT by neonates was not evidence of an innate, *distinctively human*, linguistic endowment.

Figure 7.4 explains both infants' and chinchillas' categorical perception of the voiced-voiceless contrast as the result of species-nonspecific dipole competition. In figure 7.4a, the left pole responds to an aperiodic plosive burst at $t = 0$ ms. Despite the brevity of the burst, feedback from $F^2$ to $F^1$ causes site $u^1$ to become persistently activated. This persistent activation also begins lateral inhibition of $v^1$ via $i_{uv}$. When the right pole is later activated at $v^0$ by the periodic inputs of the vowel (voice onset at $t > 25$ ms), inhibition has already been established. Because $v^1$ cannot fire, $v^2$ cannot fire. Only the unvoiced percept from $u^2$ occurs at $F^2$.

In figure 7.4b, on the other hand, voice onset occurs at $t < 25$ ms. In this case, $v^1$ reaches threshold, fires, and establishes feedback to itself via $v^2$ *before* $i_{uv}$ can inhibit $v^1$. Now, driven by both $v^2$–$v^1$ feedback and $v^0$–$v^1$ feedforward inputs, $i_{vu}$ can inhibit $u^1$, and a voiced percept results at $v^2$.

Figure 7.4 also explains more subtle aspects of English voicing. For example, the [t] in *step* is perceived as an unvoiced consonant, but acoustically, this [t] is more like a /d/: it is never aspirated, and its following VOT is usually less than 25 ms. How then is it perceived as a /t/? In this case, figure 7.4 suggests

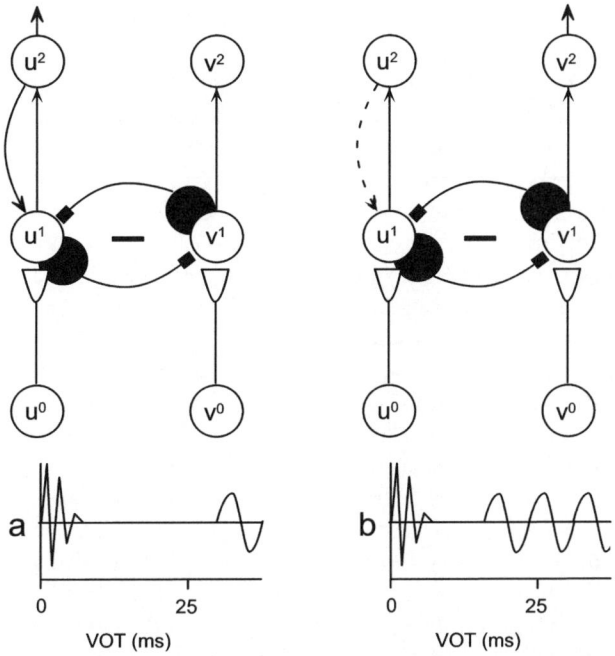

**Figure 7.4.** A VOT dipole. (a) Unvoiced percept. (b) Voiced percept.

that the preceding /s/ segment excites the unvoiced pole of figure 7.4, so it can establish persistent inhibition of the voiced pole without 25 ms of silence. It predicts that if the preceding /s/ segment is synthetically shortened to less than 25 ms, the [t] segment will then be heard as a voiced /d/.

The differentiation of wideband and narrowband perception has a plausible macroanatomy and a plausible evolutionary explanation. Figure 7.4 models a cerebral dipole, but dipoles also exist ubiquitously in the thalamus and other subcerebral structures—wherever inhibitory interneurons occur. In this case, the unvoiced pole must respond to a brief burst stimulus with a broadband spectrum. It is therefore plausible to associate $u^0$ with the octopus cells of the cochlear nucleus since, as we saw in chapter 6, this is exactly the type of signal to which they respond. Similarly, we associate $v^0$ with the tonotopic spherical cells of the cochlear nucleus. We associate $F^1$ of figure 7.4 with the inferior colliculus and medial geniculate nucleus. It is known that the octopus cells and spherical cells send separate pathways to these subcortical structures. Lateral competition between these pathways at $F^1$ is more speculative. The inferior colliculus has been mostly studied as a site computing interaural timing and sound localization (Hattori and Suga 1997). However, both excitatory and inhibitory cell types are present, and it is probable that such lateral inhibition does occur at the inferior colliculus (Pierson and Snyder-Keller 1994). Lateral competition is a well-established process at the levels of the medial geniculate nucleus, the thalamic reticular formation, and cerebral cortex (grouped as $F^2$ in figure

7.4; Suga et al. 1997). For simplicity, however, we diagram lateral competition only at $F^1$ in figure 7.4. Likewise, reciprocal feedforward-feedback loops like $u^1-u^2-u^1$ and $v^1-v^2-v^1$ are found at all levels of auditory pathways, but figure 7.4 emphasizes feedback loops from $u^2$ to $u^1$ following Suga and his colleagues (Ohlemiller et al. 1996; Yan and Suga 1996; Zhang et al. 1997), who have identified what is presumably homologous "FM" feedback circuitry from cerebrum to inferior colliculus in bats. Finally, at typical central nervous system (CNS) signal velocities of 1 mm/ms, note that the circuits of figure 7.4 are also reasonably scaled for categorical perception centered around a VOT of 25 ms.

Some languages, like Thai and Bengali, map prevoiced (VOT ≤ 25ms), voiced (VOT ≈ 0 ms), and unvoiced plosive phones (VOT > 25 ms) to *three* different phonemic categories, and replications of the Eimas study using this wider range of stimuli suggest that neonates can also perceive VOT in three categories. Nevertheless, most languages, including the European languages, divide the VOT continuum into only two phonemic categories: voiced and unvoiced. The problem is that these languages divide the continuum in different places, so before the sound spectrograph came along, this situation confused even trained linguists. For example, whereas English and Chinese locate the voicing crossover at 25 ms, so that "voiced" /b/ < 25 ms < "unvoiced" /p/, Spanish locates its voicing crossover at 0 ms., so that "voiced" /b/ < 0 ms < "unvoiced" /p/. That is, Spanish /b/ is *prevoiced*, as in figure 7.1.

As one result, when Spanish and Portuguese missionaries first described the Chinese language, they said it lacked voiced consonants, but if these same missionaries had gone to England instead, they might well have said that English lacked voiced consonants. Subsequently, this Hispanic description of Chinese became adopted even by English linguists. In the Wade-Giles system for writing Chinese in Roman letters, /di/ was written *ti* and /ti/ was written *t'i*, and generations of English learners of Chinese have learned to mispronounce Chinese accordingly, even though standard English orthography and pronunciation would have captured the Chinese voiced/voiceless distinction perfectly.

Because of its species nonspecificity, subcerebral origins, and simple dipole mechanics, categorical VOT perception probably developed quite early in vertebrate phylogeny. It is quite easy to imagine that the ability to discriminate between a narrowband, periodic birdsong and the wideband, aperiodic snapping noise of a predator stepping on a twig had survival value even before the evolution of mammalian life. Still, it is not perfectly clear how figure 7.4 applies to the perception of Spanish or Bengali. As surely as there are octopus cells in the cochlear nucleus, the XOR information of the voicing dipole is present, but how it is used remains an issue for further research.

## Phoneme Learning by Vowel Polypoles

A few speech features such as VOT may be determined by subcortical processing, but most speech and language features must be processed as higher cognitive functions. Most of these features begin to be processed in primary auditory cortex, where projections from the medial geniculate nucleus erupt into the temporal lobe of the cerebrum. Although a few of these projections might be random or diffuse signal pathways, a large and significant number are coherently organized into tonotopic maps. That is, these projections are spatially organized so as to preserve the frequency ordering of sound sensation that was first encoded at the cochlea.

PET scans and MRI scans have so far lacked sufficient detail to study human tonotopic maps, so direct evidence of tonotopic organization in humans is sparse. Animal studies of bats and primates, however, have revealed that the typical mammalian brain contains, not one, but many tonotopic maps. The bat, for example, exhibits as many as five or six such maps (Suga 1990).

It is rather impressive that this tonotopic order is maintained all the way from the cochlea to the cerebrum, for although this distance is only a few centimeters, some half-dozen midbrain synapses may be involved along some half-dozen distinct pathways. Moreover, no fewer than three of these pathways cross hemispheres, yet all the signals reach the medial geniculate nucleus more or less in synchrony and project from there into the primary auditory cortex, still maintaining tonotopic organization. In humans, tonotopic organization implies that the formant patterns of vowels, which are produced in the vocal tract and recorded at the cochlea, are faithfully reproduced in the cerebrum. The general structure of these formant patterns was presented in chapter 5. What the cerebrum does with tonotopic formant patterns is our next topic.

To model how phones like [i] become phonemes like /i/, we return to the on-center off-surround anatomy, which we now call a *polypole* for simplicity. For concreteness, imagine an infant learning Spanish (which has a simpler vowel system than English), and consider how the formant pattern of an [i] is projected from the cochlea onto the polypoles of primary auditory cortex ($A^1$) and tertiary auditory cortex ($A^3$) in figure 7.5. (We will discuss $A^2$ at the end of this chapter.) If we take the infant's primary auditory cortex to be a *tabula rosa* at birth, then its vector of cortical long-term memory traces, drawn in figure 7.5 as modifiable synaptic knobs at $A^3$, is essentially uniform. That is, $z_1 = z_2 = \ldots = z_n$.

When the vowel [i] is sensed at the cochlea and presented to polypoles $A^1$ and $A^3$, the formants of the [i] map themselves onto the long-term memory traces $z_i$ between $A^1$ and $A^3$.

## Feature Filling and Phonemic Normalization

In figure 7.5, lateral inhibition across $A^1$ and $A^3$ contrast-enhances the phoneme /i/. Thus, the formant peaks at $A^3$ become more exaggerated and better defined than the original input pattern. This has the benefit of allowing learned, expectancy feedback signals from the idealized phoneme pattern across $A^3$ to deform the various [i]s of different speakers to a common (and, thus, *phonemic*) pattern for categorical matching and recognition.

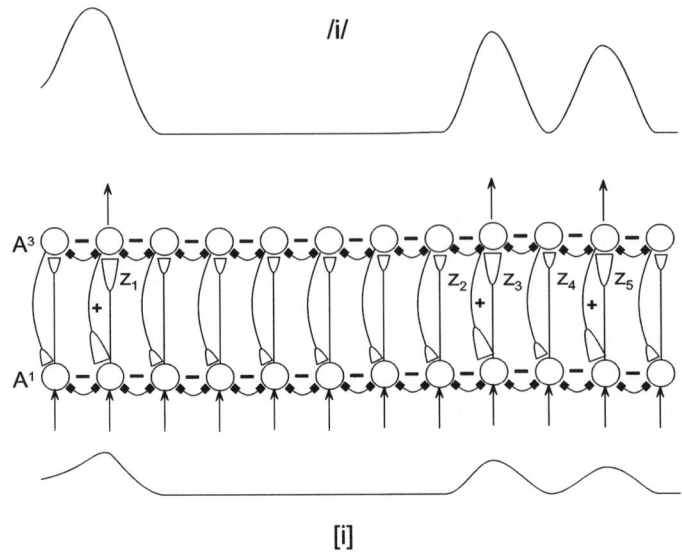

**Figure 7.5.** Propagation of [i] from cochlea to primary ($A^1$) and tertiary ($A^3$) auditory neocortex.

For example, in figure 7.6, a variant acoustic pattern, [ɪ], is input to the mature system after it has already learned the normal phoneme /i/ at $A^3$. Even though this new, variant input has its $F_2$ peak at $x_2$, feedback is stronger around the learned $A^1$–$A^3$ loop at $x_3$ because of the learned and heavily weighted long-term memory trace at $z_3$. Thus, activity at $x_3$ becomes greater than at $x_2$, even though the phonetic input to $x_2$ is greater than the input to $x_3$. As a result, the phone [ɪ] is deformed and perceived in the phonemic category /i/.

Elsewhere in psychology, this deforming-to-match is known as "feature filling." By the same kind of on-center off-surround polypole mechanisms, we can recognize a partially masked face or read a poor-quality photocopy of a document, filling in missing features from long-term memory. In all such cases, information that is missing in the sensory input is reconstructed from memory. In ART, these "top-down" signals shaped by (learned) long-term memory traces are the same expectancies we discussed in chapter 5.

## Perceptual Interference and Learning New Patterns

The ability to fill in features has obvious survival value. In certain circumstances, however, it can give rise to the phenomenon known as *interference*. For example, in Spanish, the [i] of figure 7.5 and the [ɪ] of figure 7.6 are *allophones*—variants that both match the Spanish phonemic category /i/. The Spanish speaker's long-term memory array at $A^3$ in figure 7.6 has learned to deform the lower second formant of [ɪ] to match the higher second formant of /i/, so she does

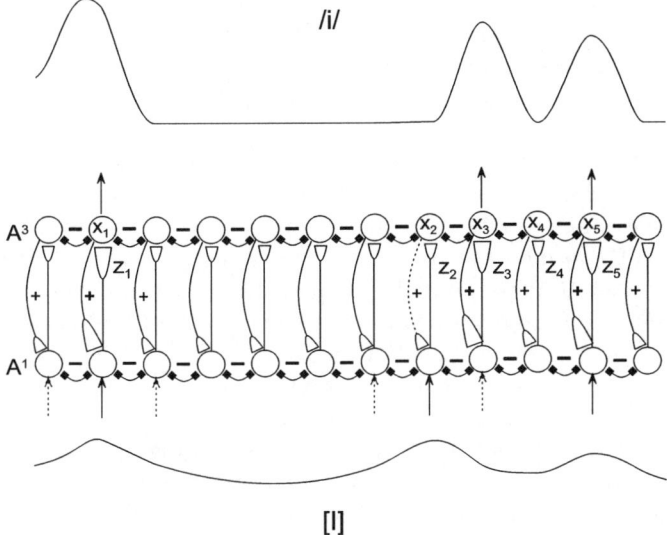

**Figure 7.6.** Phonemic normalization.

not distinguish between *bit* and *beet*.[1] This learned equivalence works fine in Spain, but when a Spanish speaker attempts to use the same circuits to process English, *interference* can arise. Like Spanish, English maps [i] onto the phoneme /i/, but unlike Spanish, English maps [ɪ] onto a distinct phoneme, /ɪ/. Thus, *beet* (/bit/) and *bit* (/b(t/) are two distinct words in English, but they would simply be different ways of pronouncing the same word in Spanish.

In the 1950s, the contrastive analysis hypothesis proposed that the more different two languages were, the more mutually difficult they would be to learn. So, for example, Italian speakers should find it easier to learn Spanish than Chinese since Italian and Spanish are both Romance languages and totally unrelated to Chinese. In general, the contrastive analysis hypothesis held up fairly well, but under close examination it was found to break down. Sometimes, second-language learners found it most difficult to learn things that were only minimally different from their first language. This interference caused a serious conceptual problem that behaviorism was unable to solve. Why should both maximally different and minimally different structures be more difficult to learn than structures that were only moderately different? Figure 7.6 explains what behaviorism could not: previously learned features only interfere with new features within their off-surround. The polypole simply resolves this contradiction between interference theory and contrastive analysis.

In the early stages of learning a second language, gross and confusing miscategorization of speech sounds is a familiar experience, but it can be relatively quickly overcome. To say that our Spanish speaker does not distinguish *beet* and *bit* is not to say that she cannot learn to do so. But *how* does she learn? If $x_3$ dominates $x_2$ in figure 7.6, how can $x_2$ ever activate in response to [ɪ]? For the answer, recall how a flash of white light rebounded a green percept to red in figure 5.9. White light accomplished this rebound because it contains all colors, including green and red, and so it stimulated both poles of the retinal dipole nonspecifically. The flash of white light was an example of nonspecific arousal (NSA).

Like the flash of white light in the red-green dipole of chapter 5, any surprising, arousing event tends to elicit NSA and so has the capacity to rebound cortical activity and initiate the learning of new information at contextually inactive sites. To understand how this learning begins, let us continue the preceding example by imagining that our Spanish speaker has been wondering at the force with which English speakers use the word *sheet*. Suddenly she realizes that they are not saying /ʃit/ at all; they are saying /ʃɪt/. This shocking development unleashes a neocortical wave of NSA, which rebounds the polypole of figure 7.6 as depicted in figure 7.7.

In figure 7.7, as in figure 7.6, the phone [ɪ] is being presented to $A^1$. In figure 7.7, however, a burst of NSA has rebounded the $A^1$ polypole. In $A^1$, $x_2$ and $x_4$, which had previously been dominated by $x_3$ and its strong long-term memory trace, have been nonspecifically aroused and have wrested control from $x_3$. Now $x_2$ is active and its long-term memory trace at $z_2$ can grow in response to the bottom-up input of [ɪ].

Bilingualism

In a polypole like $A^3$ of figure 7.7, NSA turns off the sites that were on and turns on the sites that were off. But now what is to prevent $z_2$ and $z_3$ in $A^3$ from equilibrating? In figure 7.8, the $A^3$ long-term memory traces encoding [ɪ] as

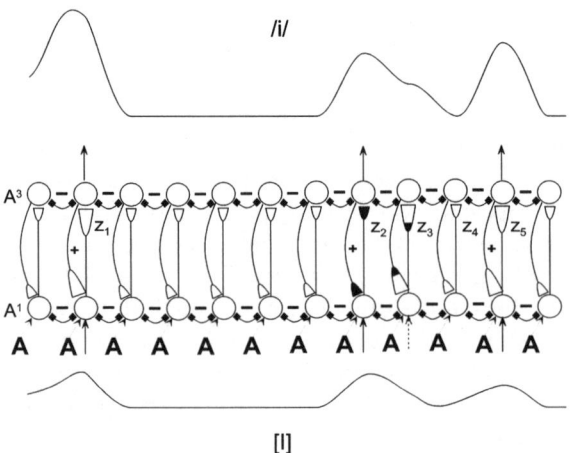

**Figure 7.7.** Rebound across the vowel polypole of /i/.

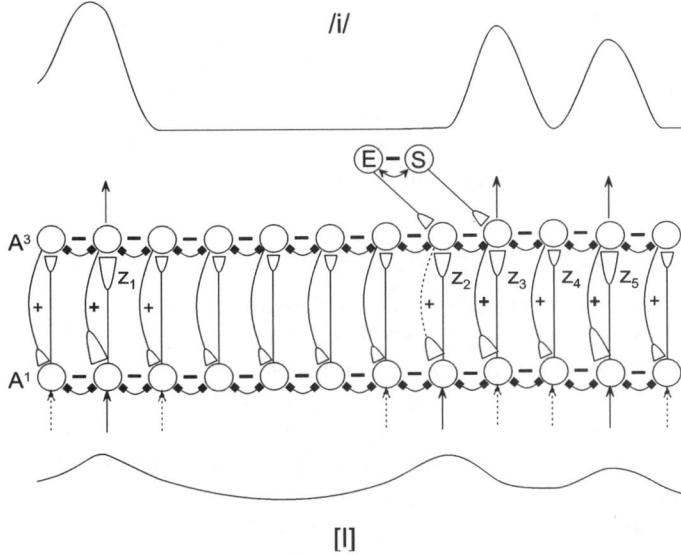

**Figure 7.8.** A dipole enables bilingual code switching.

/i/ (Spanish) and /ɪ/ (English) have reached equilibrium. In this state, how could such a "balanced" bilingual ever know which language is being spoken? Why don't balanced bilinguals randomly perceive [ɪ] as either /i/ or /ɪ/ (or, mutatis mutandis, produce /ɪ/ as either [i] or [ɪ])? Worse, how can the balance be maintained? If such a bilingual moves to a community where Spanish is never spoken, why doesn't he or she forget Spanish promptly and utterly? Why are learning and unlearning not strictly governed by overall input frequency?[2]

One answer, which we discovered first in chapter 5, is that a rebound complements memory into active and inactive sites, so that the new input becomes remembered at sites which have been inactive in the current context. No brain cell is ever activated without activating other neurons, so these inactive sites encode the new memory in a new, contextually modulated subnetwork. So our answer to the long-term invariance of language learning lies in a contextual Spanish-English dipole like that of figure 7.8. When the balanced bilingual is in a Spanish context, the Spanish pole of the contextual subnetwork is active. This biases the $A^3$ polypole toward interpreting [ɪ] as /i/. But when the balanced bilingual is in an English context, the English pole of the dipole is active, and the phonemicization network is biased toward the English phoneme /ɪ/. In mixed contexts, the dipole can oscillate, and the balanced bilingual can "code-switch" in centiseconds between Spanish and English and between /ɪ/ and /i/.

This code-switching dipole was predicted in Loritz 1990. In 1994, Klein et al. reported a PET study of bilinguals that appears to have located part of just such a dipole in the left putamen. They analyzed cerebral blood flow when sequential bilinguals, who had learned a second language after age five, repeated words in both languages. There was a significant increase in blood flow in the left putamen when the second language was spoken. Considering that the putamen and the other basal ganglia are also implicated in parkinsonism, a disorder of tonic muscular control, a plausible hypothesis is that the left putamen is implicated in maintaining articulatory posture.[3]

Vowel normalization

In the last chapter we observed that, because vocal tracts are all of different lengths, the infant language learner faces the daunting task of phonemic normalization. That is, in the clear case of vowels, how is a child to learn that [$i_{mommy}$], [$i_{daddy}$], and [$i_{baby}$] are all allophones of /i/ when mommy, daddy, and baby all have vocal tracts of different lengths and therefore vowel formants at different frequencies? Yet Kuhl (1983) established that infants learn that mommy's and daddy's vowel sounds are equivalent in the first year of life!

The first part of the answer is to be found in a classic study by Peterson and Barney (1952). They asked seventy-six men, women, and children to record the vowels [hid], [hɪd], [hɪd], etc., and spectrographically measured their formant values. The results are presented in figure 7.9.

The various English vowels clustered along axes in a 2–space defined by $F_1$ and $F_2$, joined at the origin. Within each phoneme, male vowels were located toward the low-frequency pole of the cluster while children's vowels were located toward the high-frequency pole, with female vowels in between. Rauschecker et al. (1995) found just such an array in $A^2$ of rhesus monkey cortex: two tonotopic maps, joined at the origin. Many monkeys have two types of calls which can be broadly classed as /i/- and /u/-calls, and these calls raise the same basic problem as human vowels. To determine if the call is the call of mommy, daddy, or child, the calls must somehow be perceptually normalized. The hypothesis that these two rhesus maps project to an $A^3$ like the Peterson and Barney vowel chart has not been tested, but logically such a process must intervene between audition and final phoneme perception in the human case. Positing such a normalization mechanism in $A^2$, I omitted $A^2$ from figures 7.5–7.8.

## Tonotopic Organization

Having now discussed the dynamics of polypoles at some length, we can return to the question first raised in chapter 5 in connection with the topographic organization of striate cortex. The existence of retinotopic organization in vision and tonotopic organization in audition lends itself naturally to the theory that the brain is genetically preprogrammed in exquisite detail. But as noted in chapter 5, with some $10^8$ rod cells in the retina alone and only $10^5$ genes in the entire genome, this interpretation had to be less than half the answer. Most of the answer apparently has to do with the on-center off-surround anatomy of the afferent visual and auditory pathways. To illustrate how polypoles en-

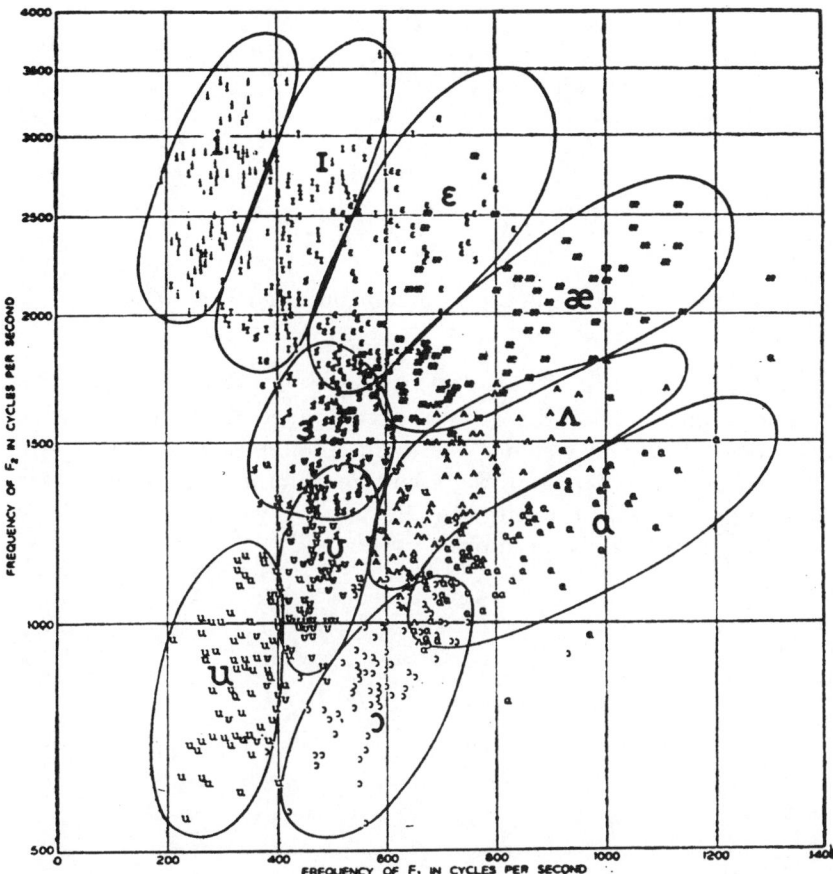

**Figure 7.9.** English vowels of male, female, and child speakers. (Peterson and Barney 1953. Reprinted by permission of the American Institute of Physics.)

force tonotopic organization, figure 7.10 follows a "F#" afferent from the cochlear keyboard to the cochlear nucleus.

In figure 7.10, four axon collaterals leave the cochlea, C, encoding the frequency $F\#$. At the cochlear nucleus (CN), three arrive at a common site ($F\#_{CN}$), but one goes astray to G. As F# is experienced repeatedly, long-term memory traces in the F#-F# pathway will develop, and at CN, $F\#$ will inhibit G. By equation 5.2, the long-term memory trace from $F\#_C$ to $G_{CN}$ will not develop. With experience, the tonotopic resolution of C-CN pathways will become contrast-enhanced and sharpened.

In this chapter, we have seen how dipoles and polypoles can account for the phonemic perception of voice onset time, the phonemic categorization of vowels, feature completion, phonemic interference, tonotopic organization, and vowel normalization. These are all low-level features of speech and audi-

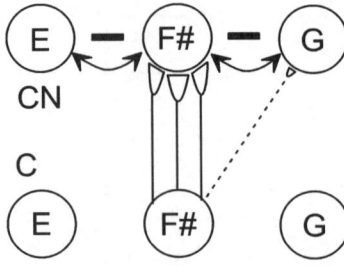

**Figure 7.10.** Tonotopic organization enforced by polypoles.

tion, and for the most part, they find analogues in the more widely studied visual system. The fact that these auditory cases are rather simpler than comparable cases in the visual system makes them a better starting point for understanding the essentials of cognitive organization. In the next chapter, however, audition finds its own complexity in the fact that speech is a serial behavior—indeed, the most complex serial behavior known.

• E I G H T •

# One, Two, Three

> Pooh and Piglet were lost. "How many pebbles are in the sock?" Pooh asked.
> "One," Piglet said.
> "Are you sure?" Pooh said. "You'd better count again, carefully."
> Piglet counted very slowly.
> "One."
>
> A. A. Milne

One, two, three, four, five, six, seven, eight, nine, ten. This seems to form a simple and perfectly natural sequence. And since the microscope had revealed that one neuron connected to the next, behaviorists were quick to fasten on the notion that these neural connections formed "stimulus-response chains." In such a chained sequence, the neuron for *one* could be thought to stimulate the neuron for *two*, which stimulated the neuron for *three*, and so on, like the crayfish tail in figure 2.6.

Although generative philosophy seemed to reject behaviorism after Chomsky's review of Skinner's *Verbal Behavior*, it did not reject behaviorism's belief that the brain is a serial processor. In a serial computer program, one machine instruction follows another. Generative philosophy and artificial-intelligence theory merely replaced the notion that one mental stimulus follows another with the notion that one mental instruction follows another. Like behaviorism, this serial theory yielded superficially satisfying initial results, but the effort ultimately failed to solve many of the same cognitive and linguistic problems that behaviorism had failed to solve.

## Bowed Serial Learning

In the first place, serial theories could not account for the fact that children, when learning to count to ten, go through a stage in which they count *one, two, three, eight, nine, ten.* Explanations invoking the child's "limited attention span" or "limited memory span" do not come to the crux of the matter. Such explanations just mask the behaviorist assumption that serial processing must underlie serial performance. The *middle* of the list gets lost. If stimulus-response chains were really responsible for such serial learning, one would expect the *end* to be forgotten. Why is it that the end is *remembered*?

Nor is learning to count an isolated case. Difficulty with the middles of lists appears ubiquitously in the experimental psychology literature under the rubric of the "bowed learning curve" (figure 8.1, see Crowder 1970 for a paradigmatic example). The bowed learning curve describes a pattern of results in which items at the beginning of a list and at the end of a list are remembered better (or learned faster) than items in the middle. But why? To understand the bowed learning curve, consider figure 8.2, which illustrates how a competitive, parallel anatomy learns to count to three.

In figure 8.2, we look more closely at how $x_j$, a node in a parallel, on-center off-surround cerebral anatomy, can learn to count to three. That is, $x_j$ must somehow faithfully remember the order of the three $x_i$ motor patterns $x_1$, $x_2$, and $x_3$, which correspond to the English words *one*, *two*, and *three*. In an ART anatomy, $x_j$ must remember this at its three long-term memory (LTM) sites, $z_{j1}$, $z_{j2}$, and $z_{j3}$.

Recall now that any $z_{ji}$ can grow only when both sites $x_i$ and $x_j$ are "on" (see table 5.1 and equation 5.2). Then, at time $t = 1$, $x_1$ is active, and $z_{j1}$ grows.[1] At $t = 2$, $x_2$ will be activated, but it will be inhibited by the persistent, lateral inhibitory surround of $x_1$. At $t = 3$, $x_3$ will be activated, but it will be inhibited by *both* $x_1$ and $x_2$, so the trace $z_{j3}$ cannot grow as rapidly as $z_{j2}$, much less $z_{j1}$. In time, with repeated rehearsal, the gradient of LTM strengths in figure 8.2 will become $z_{j1} > z_{j2} > z_{j3}$, and $x_j$ will remember the serial order *one, two, three*. Thereafter, activation of $x_j$ will cause the remembered serial pattern to be "read out" across $x_{1-3}$: $x_1$ will be gated by the largest LTM trace, $z_{j1}$, so it will receive the largest signal from $x_j$. The first motor control site to reach threshold will therefore be $x_1$ and the sys-

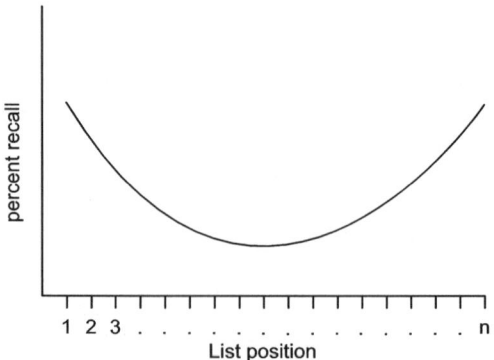

**Figure 8.1.** Bowed learning curve.

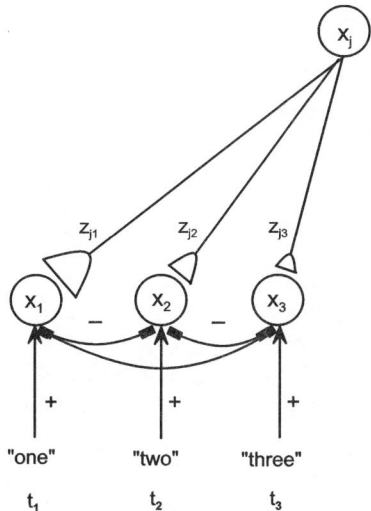

**Figure 8.2.** Learning to count to three.

tem will perform the word *one*. After $x_1$ is produced, $x_2$, gated by the next-largest LTM trace, $z_{j2}$, will be the next site to reach threshold, and it will perform *two*. Finally, $x_3$ will perform *three*. In this manner, the serial behavior *one, two, three* is learned and performed by a parallel, cerebral architecture.

There are, however, problems and limits to this simple parallel architecture. Figure 8.3 depicts the first such problem, which is encountered when learning the end of a list. In figure 8.3, when *nine* is learned, $x_9$ inhibits the next item at $x_{10}$. But inhibition from $x_9$ also works backward, inhibiting $x_8$! When $x_{10}$ is learned, it will likewise inhibit $x_9$ and $x_8$. But if $x_{10}$ is the last element of the list, there will be no $x_{11}$ or $x_{12}$ to inhibit it! Accordingly, an $x_8 < x_9 < x_{10}$ short-term memory (STM) activity gradient will develop. With time, this will translate into a $z_{j8} < z_{j9} < z_{j10}$ LTM gradient.

This kind of backward learning defied explanation under serial theories, but ART still has some explaining to do, too. Otherwise, it would imply that children learn to count backward when they learn to count forward! Before addressing this problem, let us see how the LTM gradients solve the problem of the lost middle.

If we combine figures 8.2 and 8.3 in figure 8.4, the LTM gradient $z_{j1} > z_{j2} > z_{j3}$ creates a "primacy effect" whereby earlier elements of a list are learned better and faster. At the same time, the LTM gradient $z_{j10} > z_{j9} > z_{j8}$ creates a "recency effect" whereby later elements of a list are learned better and faster. The middle of the list is inhibited by both of these effects. That is why it is learned worst and last.

So why don't children learn to count to ten in the fashion of *one, two, three, ten, nine, eight*? In order to completely account for serial learning, we must first

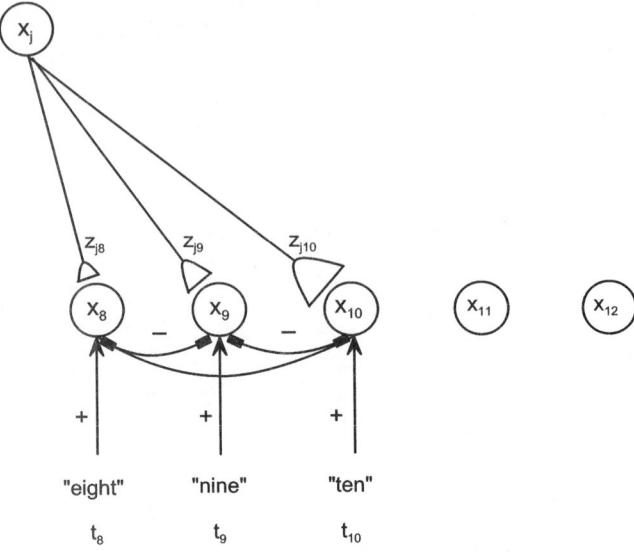

**Figure 8.3.** Learning eight, nine, ten.

differentiate between short and long lists. Short lists like *one, two, three* can hardly be said to have a middle. They tend to exhibit primacy effects and are not prone to bowing and recency effects. These latter effects only begin to appear in longer lists.[2] To achieve a reliable performance, the child must "chunk" this long list into several shorter sublists, each organized by the primacy effect, for example, (*one two three four*) (*five six seven*) (*eight nine ten*).

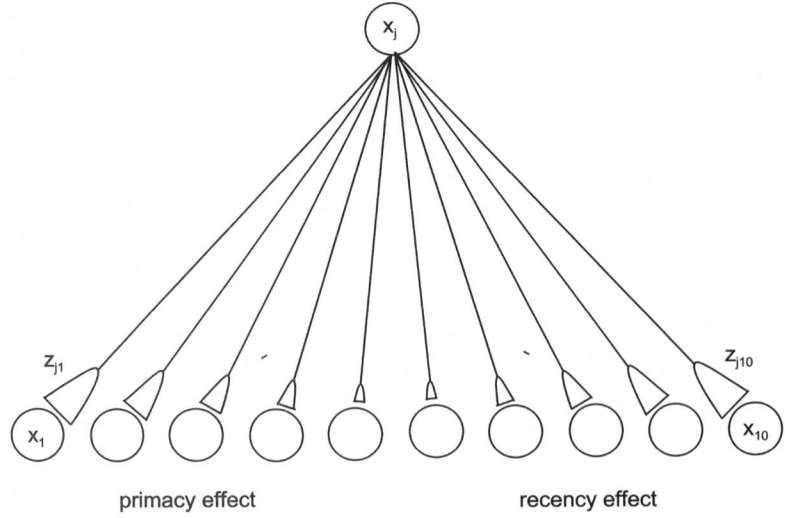

**Figure 8.4.** Primacy and recency effects produce bowing.

## Unitization

In a famous paper, "The Magic Number Seven," George Miller (1956) reviewed the serial-learning literature and concluded that seven, "plus or minus two," was an apparent limit on the length of lists which could be learned. Miller argued that any longer list would normally be "chunked" and memorized as a list of several smaller sublists. Note, however, that even lists of length seven, like U.S. phone numbers, tend to be broken into smaller chunks of three or four items.

From figures 8.2–8.4, we can observe that serial bowing depends largely upon the extent of the inhibitory surround. For example, if the radius of inhibition in figure 8.4 were only two nodes, then the absence of $x_{11}$ would cause a recency effect to appear at $x_9$. If, however, the inhibitory surround extended three nodes left and right, the absence of $x_{11}$ would create a recency effect at $x_8$. Accordingly, we may take the extent of inhibitory axons to provide a physiological basis for the "magic number." A typical, inhibitory, cortical basket cell axon collateral might have a radial extent of 0.5 mm and synapse with some 300 target neurons (Douglas and Martin 1990).[3] Along a single polypole radius within 5 degrees of arc, a single collateral would therefore synapse with about 4 target neurons, making four items a reasonable biological upper limit on the transient memory span. We therefore take the magic number to be more on the order of "four, plus or minus two." Following Grossberg, we will call this the *transient*, or *immediate*, memory span, and we will refer to the "chunking" process as *unitization*.

## Perseveration

The preceding discussion explains how serial behavior like *one, two, three* can be learned by a parallel brain, but it raises yet another critical question. Since *one* is performed first because it dominates *two* and succeeding items, why doesn't *one* tyrannically maintain that domination? Why doesn't the anatomy perseverate and count *one one one one one* . . . ? In fact, this is very nearly what happens when one stutters, but why don't we stutter all the time?

Following Cohen and Grossberg (1986), we can solve this problem by simply attaching an inhibitory feedback loop to each node in figure 8.4, as in figure 8.5. Now, when *one* completes its performance, it inhibits itself, thereby allowing *two* to take the stage.

This is a simple solution to the stuttering problem, but it is not without its own complications. The inhibitory feedback loops in figure 8.5 are "suicide loops." If they inhibit the $x_i$ sites as soon as the $x_i$ are stimulated, then no learning could ever occur! Each $x_i$ would also immediately cease inhibiting its neighbors, and no serial order gradient could be learned either!

Grossberg (1986) suggested that an (inhibitory) "rehearsal wave" could turn these suicide loops off while the system was learning. *One* would thereby be allowed to perseverate during learning and so inhibit *two, three*, etc., long

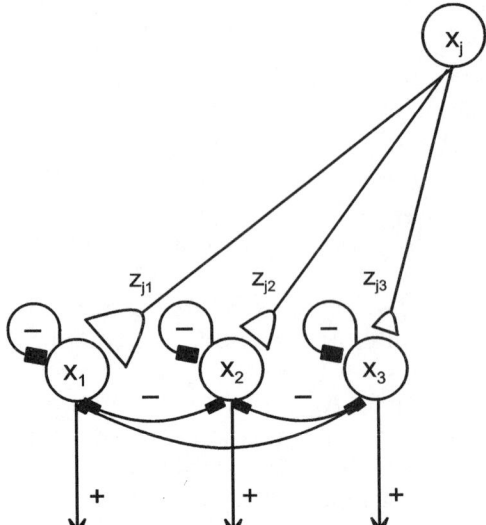

**Figure 8.5.** Anatomy for serial learning with self-inhibition.

enough for the seriating STM-LTM gradient (figure 8.4) to develop. Later, during performance, the rehearsal wave could be turned off and "suicide" inhibition enabled to prevent perseveration. The problem with this solution is in finding the suicide loops. If they are endogenous to the cerebrum, they violate the principle of on-center off-surround cerebral organization. There is, however, another way to draw the suicide loops of figure 8.5, and it involves the cerebellum.

In the beginning of a serial performance, assume frontal neocortex forms a broad motor plan. In sentence 8.1, for example, it may be assumed to plan a word like *runs*.

The dog runs down the street. (8.1)

We assume this because in Broca's aphasia, inflections like the *-s* of *runs* often fail to form. Similarly, in dysarthrias, motor disabilities affecting speech, which may result from lesions near Broca's area, the individual phonemes of a word like *run* may also be misplanned. By this analysis, the cerebrum passes a sequence of commands such as ([r]-[ʌ]-[n])-[z] to the terminal articulatory musculature. En route, however, these signals are modulated by the cerebellum, whose task it is to coordinate such fine motor activity.

Several things happen in the schematic cerebral-cerebellar circuit of figure 8.6. Descending cerebral motor commands are passed down the pyramidal tract (PT), but axon collaterals carry copies of these commands to brain stem relay points like the *pontine nucleus* (PN) and the *inferior olivary complex*

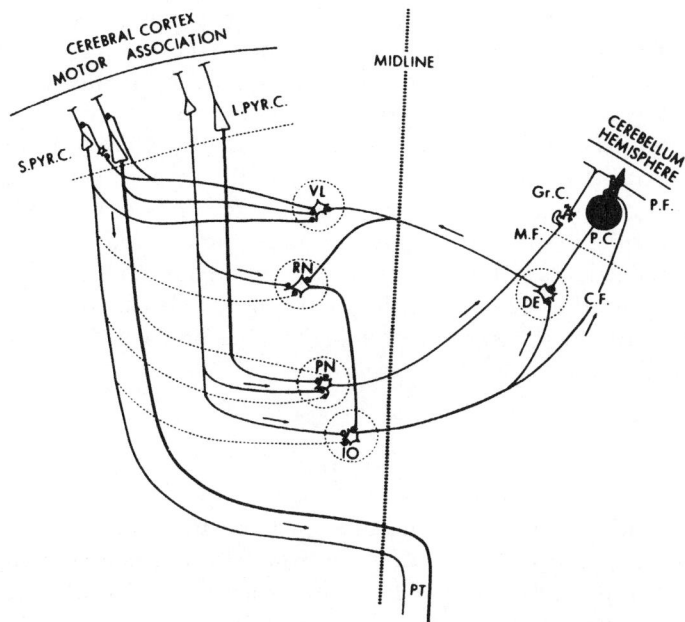

**Figure 8.6.** Cerebellar modulation and feedback. (Allen and Tsukahara 1974. Reprinted by permission of the American Physiological Association.)

(IO), where the descending motor commands are joined by ascending, *proprioceptive* signals. A loop can be traced through these nuclei, across the midline to the *dentate nucleus* (DE) of the cerebellum, and back to motor cortex through the *ventrolateral thalamus* (VL). The signals in this loop are excitatory and as such generate on-center feedback. By itself, this loop could cause the cerebral motor command to perseverate: in our example of *runs*, it might produce *r-r-r-r.* . . .

Mossy fibers from the pontine nucleus and climbing fibers from the inferior olive cross the midline and project directly to the cerebellar hemispheres. (Unlike the cerebrum, the right half of the cerebellum controls the right half of the body.) When articulation of [r] is sufficiently accomplished for articulation of [ʌ] to begin, the cerebellum signals this fact to the dentate nucleus. Now it happens that the main, large, output neurons of the cerebellum, the Purkinje cells (P.C.), are inhibitory. This means cerebellar output serves to break the perseveratory on-center feedback loop. Purkinje cells are our "suicide cells."[4] (The reader should experience a certain sense of déjà vu in this account. The circuitry of figure 8.6 is essentially the same circuitry as the six-celled brain we evolved in figure 2.9, and it solves essentially the same problem: how to stop a motor command. A car needs brakes as much as it needs an engine.)

Now let us return from the issue of perseveration in serial performance to look again at the serial-learning process described in figures 8.5. If cerebellar inhibition is active when *one* is presented during learning, then $x_1$ will not inhibit $x_2$, $z_1$ will not become greater than $z_2$, and serial order will not be learned! Clearly, cerebellar inhibition must be shut *off* during learning, but turned *on* during performance. This can be done in several ways.

First, cerebellar inhibition can be overridden by competition from cerebral excitation, that is, by attention. For example, when learning to play a B-major arpeggio, one might visually augment the nascent B-D#-F#-B motor plan by looking at a score. Simultaneously, one might repeat these note names aloud. Such additional inputs could override cerebellar inhibition during learning. Later, these supplemental excitatory inputs can be turned off for an "automatic" performance. A second available mechanism, which does not depend upon such cognitive crutches, is *tempo*.

## Tempo

The tempo with which any serial behavior is performed, be it a word or an arpeggio at the piano, can be broadly controlled by nonspecific arousal (NSA). As we generally increase nonspecific inputs to an ordered gradient like the $x_i$'s in figure 8.5, each site reaches threshold and fires sooner, with the result that the entire sequence is performed faster. At slower tempi, $x_i$ has more time to inhibit $x_{i+1}$. Thus, slow tempi maximize the slope of the serial order gradient during learning.

The climbing fibers which arise from the inferior olivary complex of the medulla to excite the Purkinje cells are one often-noted source of tempo signals in the cerebellum. It is unclear to what extent this set of inputs can be brought under conscious control or otherwise manipulated during learning, but as the tempo of these nonspecific inputs decreases, excitation of the Purkinje cells will decrease and cerebellar inhibition of the cerebral motor plan will decrease. Thus, slow tempi can also increase the LTM gradient across the $x_i$ field during learning.

It seems normal that a combination of the foregoing mechanisms operates when one learns to play a B-major arpeggio. At first the motor plan is practiced very slowly. This allows $x_i$ sites corresponding to B-D#-F#-B to achieve a steep order gradient. Eventually, the gradient is copied into LTM, and the pianist no longer has to look at the music or say "B-D#-F#." It is then sufficient to simply activate the "arpeggio array" $x_j$, and the appropriate fingering is elicited "automatically." As the tempo increases, the cerebellum activates more quickly and keeps the fingering of the arpeggio coordinated by deperseverating each note more quickly.

## Stuttering

A similar analysis can be applied to the learning and fluent performance of the serial phonemes of a word. But if during performance the cerebral tempo is much faster than the cerebellar tempo, the cerebellum cannot inhibit the cerebral motor commands fast enough or strongly enough, so the commands perseverate and the speaker stutters. A second kind of stuttering might arise if cerebellar inhibition is too strong or the cerebellar tempo too fast. In this case, the motor plan can never get started.

The preceding model is also convergent with data on ataxic dysarthria due to cerebellar disease (Schoenle and Groene 1993).[5] In particular, Kent et al. (1979) spectrographically examined five subjects with degenerative cerebellar disease. Some 50% of the phonetic segments they measured exceeded normal durations by more than two standard deviations. Figure 8.7 displays such lengthening for [p] and [k].

It is normal for speech to be slowed by a wide range of neurological disorders which affect language. In the trivial case, the afflicted speaker simply slows down in a conscious response to his own difficulty in speaking. But unlike other syndromes, cerebellar ataxic dysarthria is especially characterized by the lengthening of normally brief segments such as unstressed vowels, lax vowels, and consonants in clusters. The durations of these features and voice onset time are not normally under conscious control, and so indicate, as Kent et al. also intimated, that the cerebellum fine-tunes motor speech performance by the termination of cerebral motor speech commands.[6]

## Dipole Rhythm Generators

Suicide loops can also provide the circuitry for dipole rhythm generators (Ellias and Grossberg 1975). Indeed, our first vertebrate brain in chapter 2 (figure 2.9) evolved a cerebellum for the very purpose of creating rhythmic movement. Vertebrate serial behavior is quintessentially rhythmic serial behavior. We can

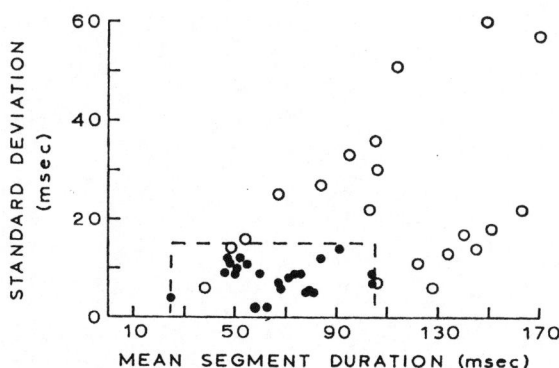

**Figure 8.7.** Lengthening of [p] and [k] by cerebellar ataxics (open circles) and normal controls. (Kent et al. 1979. Reprinted by permission of the American Speech-Language-Hearing Association.)

think of the cerebellum as the "master suicide loop" and the olivocerebellar circuit as a kind of "master clock" so long as we recognize that there are actually many suicide loops, many clocks, and many rhythms in the vertebrate brain and in the vertebrate body. Among these are the heart rhythm, the respiratory rhythm, the circadian (sleeping/waking) rhythm, the walking rhythm, and the rhythms of language, to which we now turn in chapter 9.

• N I N E •

# Romiet and Juleo

> For never was a tale of more woe,
> Than this of Romiet and her Juleo.

In chapter 8 we saw how serial order could be stored and retrieved by a parallel brain. We noted, however, that there were limits upon serial learning and performance. In particular, we noted that series of more than four items seem to exceed our immediate memory span, forcing us to *unitize* long lists as a list of sublists.

In music, the number of beats per measure rarely exceeds four. When it does so, as in a jig or a slip jig, it is usually unitized into two or three subgroups of three beats each. The same is true of English. A word like *recíprocate* has four syllables. But as soon as we go to five syllables per word, English words divide themselves with a secondary "downbeat." For example, when *recíprocate* (four syllables) becomes *rècipròcity* (five syllables), a second downbeat appears. To make matters more complicated still, the downbeats "move" to different syllables, and even the sounds of the vowels within the syllables change. In music, it is also true that measures of four beats commonly subdivide into two groups of two beats. Similarly, most English words of four syllables (and many of three syllables) also divide themselves into two beats. Thus *óperate* (three syllables) becomes *òperátion* (four syllables, two beats).

In 1968, Chomsky and Halle published *The Sound Pattern of English. SPE* was a remarkable book insofar as it brought considerable order to previously confused accounts of English stress patterns. It accounted for stress alternations like *reciprocate-reciprocity* by postulating an underlying, abstract, lexical representation in which the vowel qualities of words like *reciprocity* were marked as either *tense* or *lax*. The stress pattern of a word could then be derived from this underlying representation of vowel qualities.

And in 1966, Brown and MacNeill published a paper entitled "The 'Tip-of-the-Tongue' Phenomenon" (TOT). They presented subjects with the definitions of unfamiliar words. When subjects found the word was "on the tip of their tongue" but not quite yet definitely identified/recalled, they were instructed to write down the successive, nearly right words which came to mind. Thus, when the definition was "southeast Asian sailing vessel" and the target word was *sampan*, the word *boat* might have been an expected TOT response. Instead, what Brown and MacNeill found was that words like *salmon* or *sump pump* were more common than words like *boat* or *sailing vessel*. That is, sound-alike words, especially words beginning with the same initial phoneme as the target, were retrieved before words of like meaning. But among sound-alike words, words with a similar stress pattern to the target were recalled most frequently of all.

*SPE* could not account for this TOT result, in which stress pattern seemed to be accessed *before* any underlying vowels. By the mid-1980s, dissatisfaction with the *SPE* account had become widespread on other grounds as well, and the description of English stress in terms of *metrical phonology* (Liberman 1979; Selkirk 1984; Pierrehumbert 1987; see Goldsmith 1993 for later developments) became widely accepted.

Our account will support the general thrust of these metrical analyses. Words have *feet*, just as poets have always claimed. These feet may usually be associated with certain segmental and featural patterns of a word, but they are not serially derived from such patterns. Rather, the stress patterns of words exist in parallel with phonetic patterns (in the later jargon of metrical phonology, they exist in parallel *tiers*). This is why stress patterns, initial segments, and meaning can all be activated independently in the TOT phenomenon. In this chapter, we will see how these "tiers" of metrical phonology are ultimately and universally based on cortical cytoarchitecture.

## Spoonerisms

As we noted in chapter 1, *spoonerisms* are named after the *mal mots* of Rev. William Archibald Spooner (1844–1930), Fellow and Warden of New College, Oxford. On one occasion, meaning to address a group of workers as "sons of toil," he supposedly instead called them "tons of soil." In point of fact, Dr. Spooner probably uttered only a handful of the many "spoonerisms" attributed to him by his Oxford undergraduates. Hardly an idiosyncratic quirk of Dr. Spooner, spoonerisms are ubiquitous and extremely easy to produce. What attracts our interest here is not that they are bizarre, but rather that they are so very natural.

Karl Lashley seems to have been the first to recognize that spoonerisms were more than just a joke. Lashley (1951) noted that spoonerisms were devastating to serial theories of behavior. To illustrate his argument, consider the movements necessary to account for Lashley's example, "To our queer, old dean":

To our d ear old qu een. (9.1)

Example 9.1 looks simple on paper, but if one truly believes that the brain processes the sounds of 9.1 like so many pieces of movable type, then how does it replace the *d* with *qu* before it gets to the *qu*? And where does it put the *d* until it comes to (*qu*)*een*, and how does it remember to insert the *d* when it gets there? And if this is easy to explain, then why aren't there spoonerisms like *9.2 or *9.3?

*dear old queen our (9.2)

*our dean old queer (9.3)

Considerable research into the errors of otherwise normal speech followed Lashley's questions (Chomsky 1972, 3; Fromkin 1973, 1977; Cutler 1982; Garrett 1993).

## Romiet and Juleo

One of my more infamous spoonerisms occurred when I was reading the part of Chorus in Shakespeare's *Romeo and Juliet*. Coming to the closing couplet, I dramatically closed the book and recited,

For never was a tale of more woe, (9.4)

Than this of Romiet and her Juleo.

Figure 9.1 models the cerebral organization by which I propose to excuse my metathetic performance. Let us call this neural system "Spooner's Circuit." In it, four major linguistic fields are identified corresponding to four levels of unitized motor plans: *phrase, word, foot,* and *syllable*. Every phrase unitizes some "magic number" of words ($n \leq 4$), and every word unitizes some magic number of feet. Every foot consists of two feet (or "beats"): a "left" foot (or "downbeat") and a "right" foot (or "offbeat"). Every beat unitizes a magic number of syllables. For simplicity, we will treat each unit as a dipole, and for concreteness, the reader may wish to imagine that these plans are located in concentric rings emanating rostrally from Broca's area. At the top level of figure 9.1, the phrases *Juliet and Romeo* and *Romeo and Juliet* exist in dipole opposition. (The name of the play is *Romeo and Juliet*, but Shakespeare ended the play with the phrase *Juliet and her Romeo*.) At the center of figure 9.1, but somewhat offstage in vivo, a dipole rhythm generator oscillates between the *left foot* and the *right foot*.

The boldface entries in tables 9.1–9.4 indicate which pole of each level of figure 9.1 is active at times $t_1$–$t_4$. At $t_1$ (table 9.1), the usual phrase, *Romeo and Juliet*, its first word, *Romeo*, and *Romeo*'s "left foot," *Rom*, were all activated. So *Rom* was output.

At $t_2$ (table 9.2), the rhythm generator shifts the system to the "right foot." Normally, this would activate the *eo* syllables of *Romeo*. But at this instant, I re-

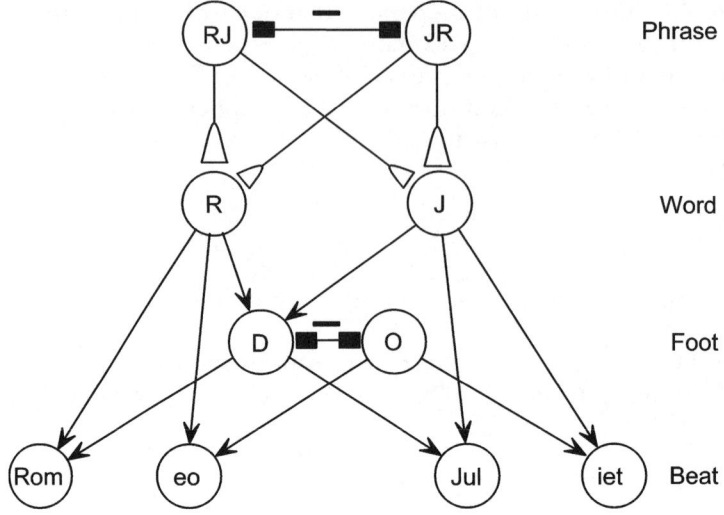

**Figure 9.1.** A Spooner circuit for *Romiet and Juleo*.

alized that my plan wasn't going to rhyme! A burst of nonspecific arousal (NSA) rebounded my phrase-level first dipole: I switched to the phrase *Juliet and Romeo*, under which *Juliet* is the first word preferred by long-term memory (LTM). But I couldn't say *Jul:* I was "on my right foot." Instead, out came *iet*.

Next, at $t_3$, (table 9.3), the system shifted to the left foot. *Juliet* was still active at the word level, so I said *Jul*.

Finally, at $t_4$, (table 9.4), the foot changed to the right foot, *Juliet* was deperseverated, and the word-level dipole rebounded. *Romeo* was selected at the word level, and *eo* was selected at the foot level. I said *eo*. And thus ended the tragedy of *Romiet and Juleo*.

## Lought and Thanguage

A greater tragedy than *Romiet and Juleo* has been the confusion of thought and language. In the spoonerism *lought and thanguage*—or in pig latin, for that matter—the metathesis cannot be driven by metrical feet in the same manner as *Romiet and Juleo*: *thought* has only one foot, and only phonemes—not entire

TABLE 9.1. Output of figure 9.1 at $t_1$.

| Phrase | Juliet Romeo | **Romeo Juliet** |
|---|---|---|
| Word | Juliet | **Romeo** |
| Foot (L) | | **Rom eo** |
| Output | "Rom . . ." | |

TABLE 9.2. Output of figure 9.1 at $t_2$, after nonspecific arousal.

| Phrase | **Juliet Romeo** | Romeo Juliet |
|---|---|---|
| Word | **Juliet** | Romeo |
| Foot (R) | Jul **iet** | |
| Output | "Romiet..." | |

syllables—are metathesized. So an additional rhythm generator must be postulated to drive this spoonerism, as in figure 9.2.

The operation of this circuit is similar to that of figure 9.1 except that at the syllable level, each syllable must be divided into an *onset* and a *rhyme* controlled by a distinct, *syllabic* rhythm generator. We presume that at $t_1$ the speaker intends to say *language*. Then for some external reason, the word dipole is rebounded at $t_2$, just after [l] is produced. As a result, the *thought* word plan is forced active, but the syllabic rhythm generator has switched from onset to rhyme: [ɔt] is output at $t_2$.

With the completion of a syllable, the foot dipole rebounds to the offbeat, and a morphological beat (*M; and*) is output at $t_3$. The foot dipole rebounds back to the downbeat, but *thought* and /θɔt/ still have not been deperseverated. With the onset pole of the syllable dipole active, [θ] is output at $t_4$.

At $t_5$, /θɔt/ and *thought* are finally deperseverated and the motor plan switches back to L. With the rhyme pole of the syllable active, [æŋ] is generated at $t_6$ (not diagrammed). At $t_7$, the second syllable of *language* becomes active. The onset pole of the syllable activates the learned serial order gradient [gʷɪdʒ]. In this and the following rhyme, figure 9.2 details the level of cerebellar deperseveration. The sounds [g] and [ʷ] are output and deperseverated under tight cerebellar control at $t_7$. No rhythmic dipole is posited at this level of the motor plan or performance, and elements at this level cannot be readily metathesized. At $t_8$, the rhyme pole becomes active and [ɪdʒ] is output in like manner.

Jakobson (1968) observed that in a remarkably large and disparate array of the world's languages, from Chinese to English, the child's first word is *mama*. Since this could not be sheer coincidence, Jakobson suggested that this universal was derived from the child's bilabial sucking reflex. I take the infant's sucking rhythm to be the prototypic syllabic rhythm generator which later subserves the organization of syllables into onsets and rhymes as well as metatheses like *lought and thanguage*.

TABLE 9.3. Output of figure 9.1 at $t_3$.

| Phrase | **Juliet Romeo** | Romeo Juliet |
|---|---|---|
| Word | **Juliet** | Romeo |
| Foot (L) | **Jul** iet | |
| Output | "Romiet and Jul..." | |

TABLE 9.4. Output of figure 9.1 at $t_4$.

| Phrase | **Juliet Romeo** | Romeo Juliet |
|---|---|---|
| Word | Juliet | **Romeo** |
| Foot (R) | Jul | Rom **eo** |
| Output | "Romiet and Juleo" | |

As figure 9.2 suggests, the lowest unitized elements of phonetic output are deactivated after performance by cerebellar deperseveration. This deactivation causes a rebound at the lowest level of cerebral planning, just as closing one's eyes can generate a McCollough effect rebound. Above the lowest levels of phonetic output, it appears that rhythmic dipoles play an increasing role, so that cerebellar deperseveration *and* dipole rebounds supply a bottom-up termination signal to motor plans.

The Spooner circuit models developed thus far not only explain metathetic "slips of the tongue" like *Romiet and Juleo* or *lought and thanguage* but also cor-

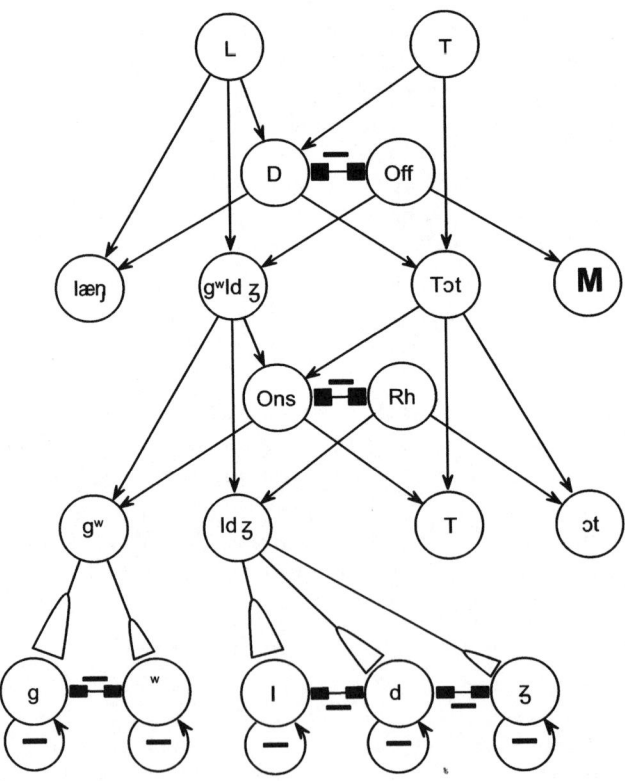

**Figure 9.2.** A Spooner circuit for *lought and thanguage.*

rectly exclude metathetic faux pas such as 9.5–9.9, which are quite unattested in the speech error literature:

*Romjul and Eoiet (9.5)

*dear our queen old (9.6)

*our dean old queer (9.7)

*langt and thouguage (9.8)

*thoughj and languat (9.9)

It appears that metrical feet are not merely a poetic metaphor. They are universal to all languages because subdividing an immediate memory span of four beats into two groups of two (or three or, rarely, four) gives greater stability to the serial plan. Perhaps it is not accidental that, after humans, the animals which we credit with the greatest ability to plan and to speak, sign, or otherwise perform serial acts with anything even approaching the complexity of language are also bipedal.[1]

In discussing metrical feet, we must be careful not to confuse basic dipole rhythm with stress, timing, or other rhythmic epiphenomena. English is often described as a "stress-timed language," apparently to distinguish its downbeat-offbeat rhythm from languages like French which do not alternate heavy and light syllables or from putatively monosyllabic languages like Chinese.[2] Downbeat/offbeat and metronomic timing may make language poetic and dancelike, but they are not necessary. Physiologically, they are quite impossible for one's feet to maintain when hiking in the woods, and musically, Gregorian chant does without them as well. Only repetitive alternation is needed for dipole rhythm.

## Offbeat Morphology

One might reasonably ask how little words like *and* or *of* affect spoonerisms and the rhythm of language. These "little words," sometimes called *functors* or *grammatical morphemes*, are the grammatical glue that holds sentences together (and in which language learners always seem to get stuck). So what role *do* the grammatical morphemes play in spoonerisms like *Romiet and her Juleo* or in *lought and thanguage?* The answer appears to be that they play no role at all.

These grammatical morphemes are universally unstressed.[3] They occur on the "offbeats" of language, and this in itself may account for a good portion of the difficulty of learning a language. In the sentence *Then the murderer killed the innocent peasant,* the nouns and verb are prominent and dramatic; it is easy to understand how such nouns and verbs can resonate in the mind and form "downbeats," milestones in the motor plan of the sentence. But what about the time-pronoun *then* or *the* or the *-ed* of *kill?* Leave them out and our sentences become telegraphic, like in Broca's aphasia. What kind of resonant subnetworks support these little words? What does "then" *mean?* The brain can't smell *then* with its olfactory bulb or see it with striate cortex or feel it with tactile cortex, so there's nothing substantive to associate *then* with in parietal cortex. It seems grammatical morphemes can exist only as auditory images in temporal cortex and motor plans in Broca's area. In what *system* can these ethereal morphemes resonate? The answer appears to be *in the grammatical system.* As critical as we must be of generative theory, it had many valuable insights, and one of them was its postulation of an *autonomous syntax.* In an adaptive grammar, this autonomous syntax seems to be more morphology than syntax and word order, but morphology does seem to organize its own networks, on the offbeat, a mortar to the substantive bricks of meaning.

From our analysis so far, we can analyze five unitization levels of rhythm and morphophonology: *phrase, word, foot, syllable, and phone sets.* At the top, as in figure 9.1, is the phrase. The phrase is a primacy gradient of substantive words. Each word is organized into feet, each with a downbeat and an offbeat. Each beat consists of one or several syllables. Each syllable can be subdivided into two phone sets: consonant(s) and vowel(s) or, more abstractly, onset and rhyme.

One can imagine how the rhythm of language might have evolved through phylogeny:

| | | |
|---|---|---|
| *Australopithecus africanus* | /ta/ | /di/ |
| *Homo habilis* | /tata/ | /didi/ |
| *Homo erectus* | /tarzæn/ | /dʒen/ |
| *Homo sapiens* | /mi tarzæn/ | /yu dʒen/ |
| *Homo loquens* | I am Tarzan. | Thou art Jane. |

These paleontological associations are fanciful, of course, but insofar as ontogeny recapitulates phylogeny, the evolutionary scheme is borne out. The child first produces single syllables and then duplicates them into two beats. Eventually, the onsets and rhymes are elaborated, and the beats differentiate themselves, forming true two-syllable words. Then the child enters the two-word stage, producing two words with the same fluency and control—in the same rhythm—as he previously produced one (Branigan 1979). In this stage, grammatical morphemes begin to make their offbeat appearance.

We were led to this rhythmic view of morphology from a more general consideration of how parallel neurons can encode serial behavior. The details of this intricate morphophonemic dance go well beyond what we can consider here, so we will return to the topic again when we consider language learning in chapter 12. But several more topics are worthy of a passing note.

In addition to *free* grammatical morphemes like *then, and,* prepositions, and the like, there are also two kinds of *bound* grammatical morphemes: *derivational* morphemes and *inflectional* morphemes.

The derivational morphemes fit easily into the rhythmic framework we have been developing. In fact, they fairly require a rhythmic analysis. For example, when an English noun like *reciprócity* or an adjective like *recíprocal* is derived from a stem like *recipro-*, the downbeat of the derived form shifts according to the suffix (Dickerson and Finney 1978). We might say that *-ity* has the underlying form *1-ity*, whereas *-al* has the underlying form *1-2-al*, where *1* and *2* are stress levels of the preceding syllables ("downbeat" and "offbeat," respectively).

A recurring research question in morphology is whether accomplished speakers of a language generate forms like *reciprócity* from such underlying representations "by rule" and "on the fly," even as they speak, or if they first learn and store *reciprócity* in its final form and then simply retrieve it, prefabricated, when they speak. This is a bit of a trick question, since it seems clear that accomplished speakers can do both. Children, however, at first seem only able to access prefabricated forms, as is shown by *wug tests*.

Berko (1958) and later Derwing and Baker (1979) assessed children's language development by measuring their ability to correctly fill in blanks like those of 9.10b and 9.11b:

This is a picture of one wug. (9.10a)

This is a picture of two *wugs*. (9.10b)

This is a picture of a pellinator. (9.11a)

A pellinator is used for *pellination*. (9.11b)

Somewhere around the age of four, as their vocabularies expand, children become able to *generate* novel inflectional forms like 9.10b in accordance with the "rules" in 9.12:[4]

+sibilant# → #Iz, *e.g., rose → roses* (9.12a)

+voiced, –sibilant# → #z, *e.g., road → roads* (9.12b)

–voiced, –sibilant# → #s, *e.g., rope → ropes* (9.12c)

Rule 9.12 says that if a sibilant phoneme (/s/, /z/, etc.) occurs at the end of a word (#), then the plural form adds *-es* (subrule 9.12a). Otherwise, the voiced or voiceless plural ending /z/ or /s/ is added. Since children of three and even two can use plural forms like *cats* and *dogs* but still flunk the *wug* test, it appears they do not generate plurals by rule. But why, at age four, should children suddenly stop saying *cats* by rote and start generating it "by rule"? An adaptive grammar would simply say they do both. Whenever an appropriate plural concept is coactivated with a [–voiced, –sibilant#] word (e.g., *cat*), a /-s/ resonance is activated. Along with that /-s/ resonance, the whole-word form /kæts/ may also be activated.[5] The case of 9.12a is, however, somewhat different. In this case, the inflectional morpheme is syllabic, and at least in later learning, it would be learned under the control of the offbeat of the metrical foot rhythm generator. Such inflections should be primarily accessible "by rule," and as we shall see in chapter 12, these inflections are especially vulnerable in language disorders like aphasia or dysphasia.

By adding dipole rhythm generators to the serial mechanisms developed in chapter 8, an adaptive grammar can easily model metathesis and related aspects of phonology and morphology. These models converge with recent theories of metrical phonology to identify rhythm as a central organizing mechanism of language. These rhythms seem to be directly relevant to the cognitive processing of (word) morphology, but in the next chapter, we will see that the model can be extended naturally to account as well for the structures of syntax.

• T E N •

# Null Movement

In chapter 9, we saw how dipole anatomies could explain phonological metathesis in spoonerisms. But as Lashley (1951) noted, metathesis is a much more general phenomenon, one that occurs in virtually all forms of serial behavior. Lashley's observations were not wasted on Chomsky, who saw that metathesis also occurred in syntax. One of the first problems Chomsky undertook to solve was the problem of relating sentence 10.1 to 10.2:

John kissed Mary. (10.1)

Mary was kissed by John. (10.2)

Although the structures of 10.1 and 10.2 are metathesized, they mean the same thing.[1] In both sentences, Mary gets kissed, and John does the kissing. Linguists say that 10.1 is in the *active* voice and 10.2 is in the *passive* voice. Chomsky recognized that if we represented 10.1 in a tree structure like figure 10.1, then the derivation of 10.2 from 10.1 could be described by simply swapping the two noun phrase (NP) nodes. He called this a *transformation*, and the earliest forms of generative syntax were accordingly known as *transformational grammar*. Of course, along the way one also must attend to myriad grammatical cleanups, like changing *kissed* to *was kissed*, but transformations on tree structures promised to describe many sentential relations, if not to solve Lashley's conundrum for all serial behavior. For example, the metathesis in 10.3–10.4 could also be explained using tree structures:

John is singing a song. (10.3)

Is John singing a song? (10.4)

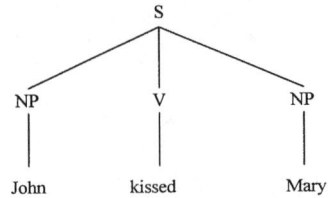

**Figure 10.1.** A basic syntactic tree.

It was the 1950s. The popular scientific metaphor was that the human brain is a computer, and Church's *lambda calculus* extension of Turing's work showed how computation could be performed on tree structures. All this implied that the brain might use a kind of lambda calculus to generate tree structures. Transformations would then operate on these tree structures to produce sentences. For example, 10.5–10.10 could be used to generate 10.11:

$$S \rightarrow NP + VP \tag{10.5}$$

$$NP \rightarrow (DET) + N \tag{10.6}$$

$$VP \rightarrow V + (NP) \tag{10.7}$$

$$DET \rightarrow a \tag{10.8}$$

$$N \rightarrow \{John, song\} \tag{10.9}$$

$$V \rightarrow is + singing \tag{10.10}$$

$$\begin{array}{l}(S\ (NP\ (N\ John)) \\ \quad (VP\ (V\ is\ singing)) \\ \qquad (NP\ (DET\ a)) \\ \qquad\quad (N\ song)))) \end{array} \tag{10.11}$$

To a computer scientist, the system of 10.5–10.10 has a number of elegant features. The NP rule (10.6) is like a computer language subroutine, a routine that is "called" by the S rule (10.5). Moreover, all the rules have the same basic structure, so they can all be computed by a single mechanism. The parenthetical syntax of 10.11 shows how the rules build on one another. This syntax may be hard to read, but if one turns the page sideways, 10.11 shows itself to actually have the tree structure of figure 10.2. Indeed, the form of 10.11 is the syntactic form of the recursive computer language LISP, which remains to this day the preferred computer language for artificial intelligence and natural language processing.

Following figure 10.2, we can model the generation of sentence 10.3 as a "top-down" process. In attempting to generate an S by rule 10.5, a recur-

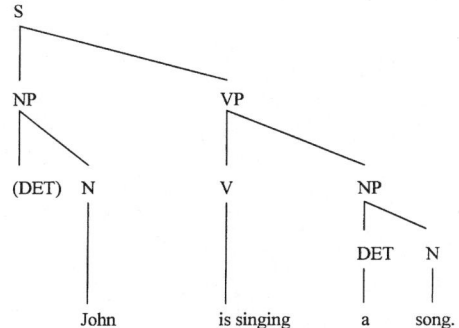

**Figure 10.2.** Tree structure of example 10.11.

sive computing device finds it must first generate an NP by rule 10.6 (without the optional DETerminer) and a VP by rule 10.7. In attempting to generate the VP, it finds it must first generate a V, and then another NP by rule 10.6. Finally, lexical insertion rules 10.8–10.10 operate to complete the sentence. In this fashion, 10.11 is built by a series of node expansions, or "rewrites":

(S (NP)        (VP)                                          )           (10.12)

(S (NP (N John)) (VP (V)        (NP)              ))          (10.13)

(S (NP (N John)) (VP (V is singing) (NP)          ))          (10.14)

(S (NP (N John)) (VP (V is singing) (NP (DET a) (N song)))))  (10.11)

To simplify accounts of movement, such as in the derivation of 10.4 from 10.3, 10.7 was later changed to move the tense of the verb out of VP, as in 10.15:

S → NP + TENSE + VP                                           (10.15)

As a result of such changes, figure 10.2 eventually came to look like figure 10.3. Now to form a question and account for 10.4, one need only move the TENSE node to the top of the tree, as in figure 10.4.

The tree structures of figures 10.1–10.4 explained a range of interesting linguistic phenomena, but perhaps the most compelling capability of the generative analysis was its ability to handle complex sentences with the same facility as simple sentences. For example, by augmenting 10.6 with a *recursive* call to the S rule, we enable the grammar to account for the problems first raised in 2.1–2.3 (repeated here as 10.17–10.19). It can account for the relative clauses in 10.17 and 10.18, and it can also block the ungrammatical sentence 10.19:

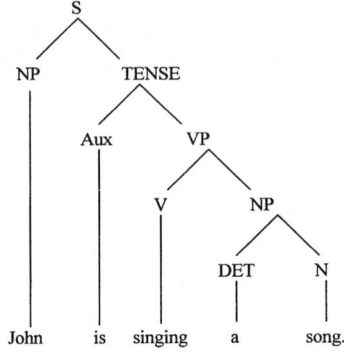

**Figure 10.3.** Lengthening the tree.

NP → (DET) + N + (S)  (10.16)

The man who is₁ dancing is₂ singing a song.  (10.17)

Is₂ the man who is₁ dancing singing a song?  (10.18)

*Is₁ the man who dancing is₂ singing a song?  (10.19)

If we ask children to make questions out of sentences like 10.17, they always produce sentences like 10.18. Why do they never move $is_1$ as in *10.19? The classic generative answer was that there are innate, universal rules of syntax, akin to 10.5–10.10, and that children's responses are governed by these innate rules.

In later years, this position became more abstract, holding only that there are innate "principles and parameters" that shape the rules that shape language, but the core computational metaphor still works remarkably well. For example, the early generative description of the clause *who is dancing* in 10.17 as an em-

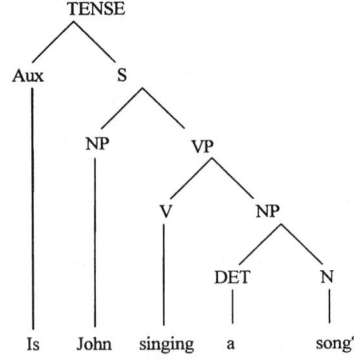

**Figure 10.4.** Syntactic tree after question transform.

*bedded sentence* led to the later postulation of a kind of *barrier* against movement between levels of recursive embedding. It is as if the parenthetic principle of 10.20 existed in the child's mind to block illegal movement as surely as it defines *dynamic scoping*[2] in the LISP computer language:[3]

($S_1$ $Is_2$ the man ($S_2$ who $is_1$ dancing) singing a song?)  (10.20)

These generative ideas brought order to a collection of linguistic data which seemed to defy explanation by any other theory. Unfortunately, as generative theory became increasingly refined, it started to look more and more like behaviorism. Unconsciously bound to the presumption that serial language must be the product of a serial processor, Chomsky found himself declaring that "the basic elements of a representation are chains"; and generative grammar's tree diagrams of the basic structure of the clause came to look more and more like stimulus-response chains and crayfish brains (Chomsky 1995; figure 10.5).

To be sure, Chomsky was not talking of stimulus-response chains, but he *was* proposing cognitive chains. Generative grammar's node-swapping insights were on the right track, but then it erred in presuming that a well-engineered human mind would node-swap the same way a computer would. With this assumption, the serial, computational, generative explanation failed to answer Lashley's criticism almost as completely as did the behavioral explanation.

In the end, generative metaphysics failed to explain *what* moves when a grammatical node "moves." We are obviously not expected to believe that when I produce sentence 10.18 in derivation from some structure like 10.4, some $is_2$ neuron physically changes its place in my brain. Generative linguists have lately taken to defending the notion of movement by claiming that it is only a "metaphor," but after forty years, science can reasonably ask what it is a metaphor *of*.

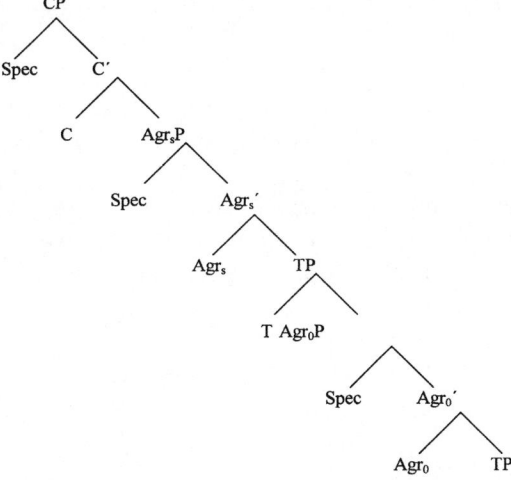

**Figure 10.5.** A syntactic tree devolves into a chain (Chomsky 1995).

If it is not neurons that are moving, then what exactly *is* moving? The simple answer, of course, is that *nothing moves*. Linguists may "derive" 10.2 from 10.1, but in normal discourse, normal humans do not. There is no such thing as linguistic *movement*. But then how *do* sentence pairs like 10.1 and 10.2 spring from the brain, and how *are* they related?

## Universal Order

Generative philosophers were right in seeking "universals" of language. Despite the existence of thousands of languages, centuries of research have yielded dozens of linguistic features that are universal or almost universal. Especially noteworthy are certain universal features of syntactic order first explicitly noted by Greenberg (1968). Greenberg classified thirty diverse languages according to the manner in which they ordered subject, object, and verb. There are six possible permutations in which these elements may be arranged: OSV, OVS, VOS, VSO, SOV, SVO. In Greenberg's study and subsequent research, the first three of these proved to be virtually nonexistent, and from the accumulating evidence, Lehmann (1978a) observed that languages exhibited a profound SO unity, differing only on the relatively minor OV/VO distinction. Furthermore, Greenberg's Universal 6 stated that: "All languages with dominant VSO order have SVO as an alternative or as the only alternative basic order" (1968, 79). Bickerton (1981) went beyond Greenberg to claim that there exists a strong, universal preference for the order SVO. In his studies of pidgin languages and creoles,[4] he posited an even deeper unity. Bickerton's reasoning was that such freshly invented languages offered better evidence for the "natural" or "universal" structures of language because they were unelaborated by linguistic artifacts of history or tradition.

From the preceding line of research, I conclude that the most basic fact which needs to be explained about language is why the subject "naturally" precedes everything else.

## Subjects

Unfortunately, the traditional grammatical term "subject" has come to mean many different things. For example, consider the following defining features of the term "subject" in three simple English sentences:

John helps Mary. (10.21)

He helped her. (10.22)

Mary is helped by him. (10.23)

The subjects of 10.21–10.23 are *John, He,* and *Mary,* respectively, but subject-hood variously implies that the subject

1. *agrees* with the verb in person and number (10.21, 10.23),
2. is in the *nominative case* (10.22),
3. is the *agent* of the sentence (10.21, 10.22),
4. is the *topic* of the sentence (10.21–10.23).

In the case of English, only assertion 4 holds for all three sentences. Contra assertion 1, *he* does not agree with the verb *helped* in any overt way in sentence 10.22. Contra assertion 2, the subject has no overt case marking in 10.21 or 10.23. And as for assertion 3, *Mary* is not the agent in 10.23; it is *him* who is doing the *helping.*

Sense 4 is also the only meaning of "subject" which holds universally across languages. Sense 1 is not universal because some languages, like Chinese, do not overtly mark agreement between subject and verb at all. In many other languages, overtly marking agreement is more the exception than the rule. English, for example, only marks agreement between present-tense indicative verbs and third-person singular subjects. Pidgin languages, almost by definition, lack agreement or grammatical case inflections. In the cases of assertion 2 and 3, we find that nominative marking and agentive marking are often mutually exclusive, as between accusative languages, which mark nominative case subjects, and unaccusative languages, which mark "subjects" in the agentive (or ergative) case.[5]

Sense 4 alone applies to 10.21–10.23, it alone applies to uninflected languages like Chinese, and it alone is the sense in which the term "subject" is relevant to our analysis of universal order. Therefore, what Greenberg's, Lehmann's, and Bickerton's data suggest for adaptive grammar is that universal subject-verb order is more accurately termed universal topic-verb order. Nevertheless, sense 4, in which the subject is regarded as the topic of discourse, has been rather the least-noted meaning of the term "subject."

## Topicality

The topic is what we are talking about. Saussure described a word as a relation between a sound or a graphic *signifier* and its referent, its *significand.* The topic's significand is usually persistently present in our cognitive environment, and its signifier is usually persistent from sentence to sentence. Neurally, this persistence takes the form of a resonance in short-term memory (STM). By the analyses of chapters 8 and 9, this means the topic should have a universal syntactic primacy effect, and according to Greenberg, Lehmann, and Bickerton, so it does.

Some of the significance of topic has been previously recognized (Li 1976). Chafe (1970), in particular, anticipated much of the importance which adaptive grammar will ascribe to it. But like "subject," the term "topic" has been so variously used as to obscure its relevance to a unified theory of language. For example, Chao (1948, 1968) and Li and Thompson (1981) refer to Chinese as a *topic-comment* language rather than a *subject-predicate* language. In so doing, however, they treated topicality as a unique feature of Chinese, obscuring its universal role.

In a different usage, both structural and generative syntacticians have used *topic* to refer only to unusual, *marked*[6] constructions such as 10.24 (from Pinker 1989):

That coat I like. (10.24)

In our view, such "topicalized" sentences do not reflect topicality so much as *change of topicality*. They are not so much *topicalized* as *topicalizing*.

Linguists, particularly sociolinguists, have also noted that *old information* (or "given" information) tends to come early in a sentence, and *new information* tends to come later. In this context, old information corresponds closely to our sense of topic, but for sociolinguists, it is the *conversation*, not the *sentence*, which is the basic unit of language. In taking this perspective, they broke away early from generative linguistics and its context-free analysis of sentences. This no-man's-land between syntax and sociolinguistics was partially bridged by the field of *pragmatics*, which used topicality relations like pronominal reference to develop an *intersentential* syntax, but the essential and universal role of topic in the simple sentence was still largely overlooked. Taken together, dissident syntacticians, sociolinguists, and pragmaticians formed a broad school of "functionalism" which perceived itself to be at theoretical odds with generative linguistics. But functionalism generally failed to relate its intersentential, discourse-level conception of topic to intrasentential syntax, and so functional linguists and generative linguists lived side by side for years, each school writing in its separate journals, neither school really understanding what the other was talking about.

Under adaptive grammar, the topic of both sentence and discourse is plainly and simply what one is talking about. At the moment a sentence is about to be spoken, the topic is neurophysiologically instantiated as that word or phrase, that motor plan which is most activated by cerebral resonance. Of course, the significand of the topic usually still exists externally, in the surrounding world, in the surrounding context, or in the surrounding discourse, but adaptive grammar's definition of topic emphasizes the internal, cognitive, STM resonance. The most active STM resonance becomes the topical signifier, the "head" element of adaptive grammar's intrasentential and intersentential syntax.

Sociolinguists speak of the tendency of old information to be expressed before new information as a rule of discourse or a property of conversation. In this sense, Grice (1975) suggested that collocutors in a conversation adhere to a tacit contract to "be relevant." Adaptive grammar sees relevance as a deeper, biological injunction to say (and do) topical things first, a "rule" which applies not just to conversations but to *everything* brains do. It is a corollary of evolution: the organism that doesn't do relevant things first simply doesn't survive.

Under adaptive grammar, what one is talking about, the *topic*, is the currently most activated referential subnetwork in neocortex. By our account of serial order in chapters 8 and 9, persistent STM activation will drive this topic subnetwork to competitive, primacy-effect prominence among competing sentence elements. Therefore, in 10.21–10.23, as in all the unmarked sentences of all the world's languages, the topic is the first propositional nominal element of a sentence. But what do we mean by *propositional*, and how does adaptive grammar describe the remaining, presumably competing elements of the sentence?

## Case Grammar

Case grammar is generally said to have originated with Fillmore (1968), but there were several precursors. European languages are generally richer in grammatical cases than English. In these languages, it has always been apparent that there is a correlation between such grammatical categories as *nominative* (subject), *accusative* (direct object), and *dative* (indirect object) and such semantic categories as *actor, patient,* and *donor/recipient*. From such relations, Tesniere ([1959] 1969) had developed *valency grammar*, and Gruber ([1965] 1976) had developed a theory of *thematic relations*. In these systems, a proposition is a clause consisting of a single verb and its most immediate (or "inner") case arguments.[7]

Unfortunately, the correlation between grammatical case and semantic case is not always close. Thus, in 10.23 (*Mary is helped by him*), *him* is in the accusative case, but *him* is still the "helper," and so semantically it fills the *actor* role. Different case grammarians also use different terms for similar semantic roles. Thus, the term "agent" is often used for "actor"; "patient" and "theme" are often used for "object"; and "source" and "goal" are often used for "donor" and "recipient." Over the years, this Tower of Babel has become a religious schism. Generative linguists have *thematic relations* while other linguists have *case grammar*, and the two churches never cite each other. My usage will loosely follow Cook 1989.

From the perspective of adaptive grammar, Fillmore made two especially significant contributions to case grammar theory. First, he claimed that semantic cases like *actor* and *patient* were the actual organizational elements of "deep structure." These are the "propositional" case arguments of a transitive verb like *kill*. *Give*, by contrast, has three such propositional arguments: an agent (the giver), a patient (the gift), and a *recipient*. In contrast to these "inner" propositional case arguments, most verbs also accept a variety of "outer" or "nonpropositional" case arguments—for example, *purpose* or *location*. These cases roles are usually optional; they are often left unspecified, presumably because they are often contextually obvious.

Second, Fillmore also claimed that this semantic deep structure was initially *unordered*. But then, forced to account for syntactic order, Fillmore suggested that for each verb in the lexicon there existed a *subject selection hierarchy*. By the subject selection hierarchy, the verb *kill* would normally select the semantic *actor*, if there were one, for its subject, as in *The* killer *killed the victim*. Otherwise, if there were no definite *killer* in the semantic context (say, in a mystery novel), *kill* would select a semantic *instrument* as subject, as in *The* poison *killed the victim*. Otherwise, if there were neither *agent* nor *instrument*, the *object* would become the subject, as in the passive sentence, *The victim was killed*.

## Topicalization and Passives

But what about 10.25, a passive sentence in which the object is selected as subject, even though there is an explicit instrument *and* an explicit agent?

> The victim was killed by Mr. Green with a candlestick. (10.25)

The subject selection hierarchy, by attempting a context-free explanation of word order, fails to explain many such commonplace sentences. If analysis of language and sentence is conducted independent of context, then it is by definition conducted without consideration of topic. But where, outside the ivory tower, are sentences analyzed without regard to topic? If we analyze real language, then topicality replaces the subject selection hierarchy as the principal ordering force in syntax.

Consider, for example, the last sentence of the following extract from Peirce 1877. The first paragraph is provided for context.

> Now, there are some people, among whom I must suppose that my reader is to be found, who, when they see that any belief of theirs is determined by any circumstance extraneous to the facts, will from that moment not merely admit in words that that belief is doubtful, but will experience a real doubt of it, so that it ceases in some degree at least to be a belief.
>
> To satisfy our doubts, (10.26a)
>
> therefore, it is necessary (10.26b)
>
> that some method be found by which (10.26c)
>
> our beliefs may be caused by nothing human, (10.26d)
> but by some external permanency—by something
> upon which our thinking has no effect. . . . Such is the method
> of science (Peirce 1877, 10f–11).

The last sentence is rather remarkable in that it contains *two* topicalizing clauses and *two* passive clauses. I call the first purpose clause of 10.26 (*to satisfy our doubts*) (*re*)*topicalizing* because purpose clauses are nonpropositional: they are more often extrasententially expressed, implied, or assumed than intrasententially expressed. In this case, the clause recapitulates the primary topic of the preceding pages: Peirce has argued that doubt nags us to epistemological action. The second clause (10.26b; *it is necessary*) recapitulates the fact that Peirce is conducting a philosophical argument, and that what follows, follows from logical necessity. I call 10.26b a topicalization because adjectives are not usually topics, but this "cleft" clause "fronts" and "focuses" the proposition of necessity.

In general, adaptive grammar sees all such fronting, focusing, and topicalizing as manifestations of an underlying *topic gradient*: reflections of how resonant each clause (and each component within each clause) is in the speaker's STM. The most resonant—and therefore most topical—element reaches threshold first and is expressed first. The topic gradient is a self-similar analogue of the primacy gradient we examined in chapter 8.

The third clause (10.26c) is a passive clause. It recapitulates the secondary topic Peirce has been addressing: the various *methods* by which people avoid or relieve nagging doubts. Of course, Peirce might have said it in the active voice:

?that everybody find some method by which (10.26c')

But *everybody* has not been particularly active in Peirce's STM or the reader's. As the very title of his paper makes clear, *belief* and *methods of fixing belief* are the active topics, and in 10.26 it is their expression which is accorded primacy.

The fourth clause (10.26d) is also a passive clause. *Beliefs* is topical, and so it becomes the *subject* of the clause. Once again, this clause would sound rather odd in active voice:

?nothing human may cause our beliefs (10.26d')

Finally, Peirce introduces the new information to which this passage turns: *some external permanency . . . the method of science.*

Admittedly 10.26c' and 10.26d' are grammatical (albeit only in the most trivial sense of the term), and there are other legitimate and illegitimate reasons for using and avoiding passives. It could be argued that Peirce's use of passives in the preceding analysis reflects no cognitive principles but is simply "stylistic." However, 10.26c and 10.26d are the first passive clauses Peirce has used in three pages, so one cannot easily argue that passive clauses are a signature of his style. Nor do other instances of passive "style" invalidate the principle of the topic gradient. It is perfectly consistent for adaptive grammar to maintain that the learned norms of scientific writing can inhibit the first-person expression of the researcher-agent in STM and demote it in the topic gradient.

For another class of passives, the subject selection hierarchy suggested that agentless passives like

Mistakes were made. (10.27)

are in the passive voice because there simply is no agent to assume the subject role. Less-gullible linguists (e.g., Bolinger 1968) take such passives to be linguistic legerdemain, deliberately concealing *who* did the defective deed. In the latter case, although the agent may be very active in the *speaker's* STM, his intent clearly is to inhibit the agent's resonance in the *listener's* STM.

## Subtopics and Dative Movement

The topic gradient accounts for many other types of linguistic "movement." In addition to passive movement, generative theory claimed that 10.29 was derived from 10.28 by *dative movement*. That is, in 10.28, the "indirect object," *Neil*, follows the direct object and is said to be in the *dative/recipient* case. In 10.28, *Neil* has "moved" to before the direct object.

| | |
|---|---|
| The police gave a break to Neil. | (10.28) |
| The police gave Neil a break. | (10.29) |

In the absence of context, there is no reason to prefer 10.28 over 10.29. But, except in narrow linguistic inquiry, language never occurs independently of context. Sentences 10.28 and 10.29 would normally follow from two different conversations:

| | |
|---|---|
| The police didn't give many breaks. (But,) | (10.30) |
| The police gave a break to Neil. | (10.28) |
| *The police gave Neil a break. | (10.29) |
| The police gave *Neil* a break. | (10.31) |

In 10.30, *the police* appears as the topic of discourse, and *break* is introduced as new information, becoming a secondary topic. In this context, 10.28 is preferred. Sentence *10.29 sounds relatively odd because it gives newer information in the conversation, *Neil*, primacy over the older information, *break(s)*. Giving *Neil* additional, *contrastive* stress resolves the oddity, but the exceptional markedness of 10.31 proves the rule: secondary topics normally have primacy over tertiary topics. Conversely, in the following context *10.28 violates the topical precedence established by 10.32:

| | |
|---|---|
| The police investigated Neil. (But,) | (10.32) |
| The police gave Neil a break. | (10.29) |
| *The police gave a break to Neil. | (10.28) |
| The police gave a *break* to Neil. | (10.33) |

Sentence 10.32 establishes *Neil* as the secondary topic, but in *10.28, the new information and tertiary topic *break* is presented as secondary. Once again, contrastive stress in 10.33 can correct this abnormality. Just as chapters 8 and 9 explained phonological seriality in terms of a primacy gradient, adaptive grammar explains syntactic seriality in terms of a topic gradient. No movement is necessary.

## Particle Movement

The same topic gradient effects can be found in the case of so-called particle movement.[8] If prior topics have not been established, 10.34 and 10.35 are equally felicitous context-free sentences:

| | |
|---|---:|
| John looked up the address. | (10.34) |
| John looked the address up. | (10.35) |

But given 10.36, a discourse context establishing the prior topic *address*, 10.35 is preferred:

| | |
|---|---:|
| The address was torn off the package, (so) | (10.36) |
| John looked the address up. | (10.35) |
| *John looked up the address. | (10.34) |

## The Nominal Topic Gradient and the Verbal Relation Gradient

The preceding several sections provide evidence that the nominal elements of a sentence are organized in a primacy-ordered topic gradient. Now we return to the question of where the *verb* fits into this order. At the end of chapter 9, we observed that morphology tends to organize on the offbeat. In figure 10.6, I extend this offbeat morphology to include the verb. The verb, with its case-marking prepositions (or postpositions in languages like Japanese), exists in a primacy- (or recency-) ordered verbal relation gradient. A rhythmic dipole generates sentences by alternating between these two gradients.

In figure 10.6, *Prof. Plum killed Mrs. White in the hall with a knife* is generated by alternately outputting elements from the nominal topic gradient and the verbal relation gradient. The $T$ (topic) pole of the topic/relation dipole is initially activated by $S$. The $T/R$ (topic/relation) dipole then oscillates in phase with the downbeat/offbeat "foot" dipole described in chapter 9. This does not mean that the $T/R$ dipole rebounds with each and every beat of the foot dipole. Several feet may occur on each pole of the $T/R$ dipole, as, for example,

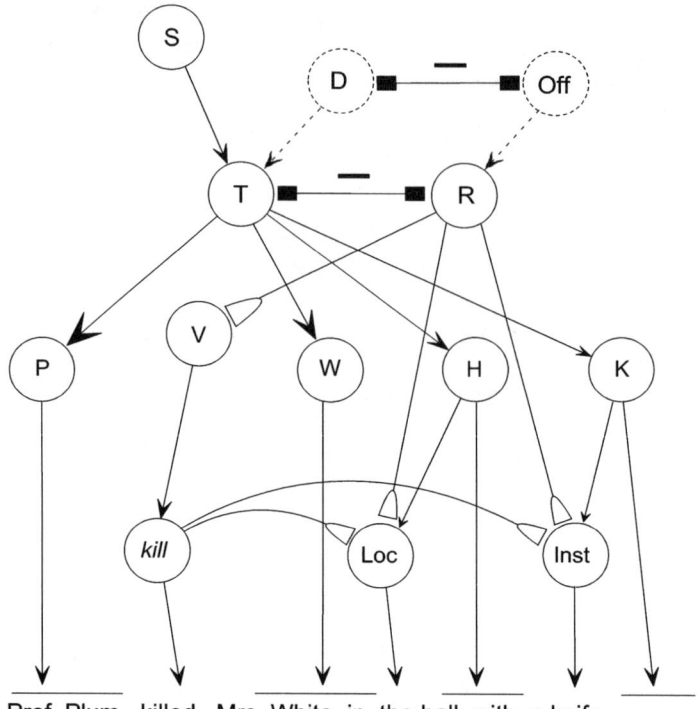

**Figure 10.6.** A nominal topic gradient and a verbal relation gradient combine rhythmically to generate a sentence.

in outputting the articles *a* and *the* or *Prof. Plum* or *Mrs. White*. All such detail is suppressed in figure 10.6 in order to more clearly illustrate the essential organizational principle of the topic/relation dipole.

The fundamental order in figure 10.6 is established by the topic gradient *P* > *W* > *H* > *K*: *Prof. Plum* is output first. (This topic gradient is contextual and transient. It is defined in STM and diagrammed in figure 10.8 with ordered STM arrows.) After cerebellar deperseveration and bottom-up rebounds have deactivated *P* and *T*, *R* rebounds into activity. *V* is then activated. *Loc* (location) and *Inst* (instrument) are "primed" (subliminally activated), because *kill* has a learned association with these case roles in long-term memory (LTM). *Killed* is output, and *V* and *R* are deactivated. The *T/R* dipole rebounds again. *Mrs. White*, the next most active nominal element in the topic gradient, is activated and output. The *T/R* dipole then rebounds again. *Loc* and *Inst* are both equally activated by *R*, since in the learned relation gradient, either could be output next, but under the current topic gradient, *H* > *K*, so *Loc* is more primed than *Inst*, and *in* is output next.

The *T/R* dipole rebounds again, and *the hall* is activated and output. The *T/R* dipole rebounds to *R*, activating *Inst* and outputting *with*. Finally, the *T/R* dipole rebounds one last time, outputting *a knife*.

## Pronouns

In chapters 8 and 9, we saw how the performed motor nodes of serial lists are deperseverated by inhibitory feedback and rebounds. This process seems also to explain fundamental universal features of pronouns, clitics, and other pronounlike words. Pronouns and related *pro-forms* are found in all natural languages, and figure 10.7 explains why sentences of the type

Sid hit himself. (10.37)

are universally preferable to sentences of the form

*Sid hit Sid. (10.38)

Figure 10.7 models pronominalization subnetworks where (a) *Sid saw Bill*, and then either (b) *Sid hit Bill* or (c) *Sid hit Sid*.

For simplicity, figure 10.7a collapses the topic and relation gradients of figure 10.6 into a single topic-relation gradient. After (*a*) *Sid saw Bill*, /bIl/ is deperseverated, so that in (*b*) the semantic relation *hit* (*Sid, Bill*) is expressed

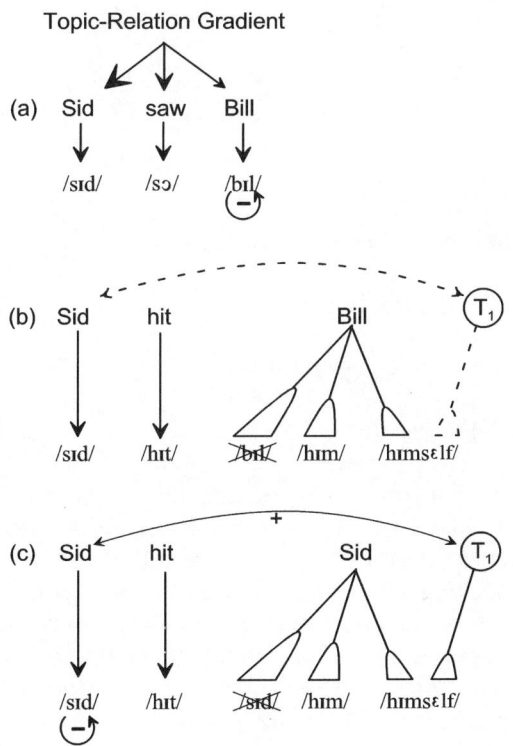

**Figure 10.7.** Simple pronominalization.

as *Sid hit him.* (By the same principles, *He hit him* is also predicted; for clarity, figure 10.7 only diagrams one pronoun.)

Figure 10.7c considers the semantic relation *hit (Sid, Sid)*. After /sId/ is initially pronounced and deperseverated, only the motor plans for /hIm/ and /hImsɛlf/ can be activated without contrastive stress. Although /hIm/ is the more frequent (and so has the larger LTM trace), /hImsɛlf/ is also activated by *T1*, the primary topic, *Sid*. By contrast, in (b), because *Bill* is not the primary topic, /hIm/ is output. There is a certain similarity between this explanation and the generative notion of *traces*.

When a linguistic element was moved, generative linguists believed that it left behind a residual *trace*. Thus, in 10.39 a trace $t_i$ of *Neil$_i$* was believed to remain in the embedded clause of the surface structure:

Neil$_i$ was believed [$t_i$ = *Neil$_i$*] to have destroyed the evidence.  (10.39)

Traces explained why, after hearing 10.39, one can reply *Neil* without any hesitation to the question *Who might have destroyed the evidence?* But since nothing moves, adaptive grammar analyzes this trace as simply a "null pronoun," the completely inhibited motor plan of its antecedent.[9]

## The Scope of Negation

Adaptive grammar also offers an explanation of the "scoping of negation." Consider 10.40, for which four interpretations (10.41–10.44) are possible:

John didn't eat the pizza quickly.  (10.40)

John didn't (NEG eat the pizza quickly).  (10.41)

John didn't (NEG *eat*) the pizza quickly.  (10.42)

John didn't eat the (NEG *pizza*) quickly.  (10.43)

John didn't eat the pizza (NEG quickly).  (10.44)

Example 10.41 interprets NEG as negating the entire scope of the verb phrase *eat the pizza quickly*, but it is more likely that *John did eat the pizza*—he just didn't eat the pizza *quickly*. Examples 10.42 and 10.43 are possible readings, but normally would be spoken with contrastive stress on the italicized words. The normally preferred specific reading is that *quickly* is being negated (10.44), and this pattern is common enough that Ross (1978) proposed a "rightmost principle of negation," which assigns negation to the final constituent of a sentence. Adaptive grammar makes a similar analysis. In 10.40, *quickly*, *eat*, and *pizza* are all activated in STM and so are potential "attachment points" for NEG. At the end of the sentence, NEG would be applied globally, presumably as a burst of nonspecific arousal, and the least-activated conceptual subnetwork, that which encodes the newest information, is rebounded.

But once NEG is encountered in the sentence, how is NSA suppressed until the end of the sentence? Is there a pushdown-store automaton in the human brain after all? And when NSA is finally released, how is it constrained so as to rebound only the rightmost element? Adaptive grammar has answers to these questions, but they are not syntactic. They must wait until chapter 12.

## Questions: Extraction and Barriers

Finally, we return to the questions raised by sentence 2.2/10.18.

Is$_2$ the man who is$_1$ dancing $p_2$ singing a song? (10.18)

Generative linguists thought the generation of 10.18 involved (*a*) the extraction of an element ($is_2$) from one place ($p_2$), (*b*) its "movement" to another place, and (*c*) an elaborate set of principled "barriers" which would, for example, prevent $is_2$ from moving to the front of the sentence. Figure 10.8 accounts for 10.18 without recourse to metaphors of movement.

English yes/no questions like 10.18 are initiated by an auxiliary verb. English *Aux* and related modal verbs carry the epistemological status of a proposition (Givón 1993). In English, this association between epistemological status (*?* in figure 10.8) and *Aux* is learned as part of the grammar, so in figure 10.8, LTM traces order *Aux* before the rest of the sentence, *S*. After *Is* is output, the *Aux-S* dipole rebounds, and *S* initiates activation of the *T/R* dipole at *T*. (The dashed LTM trace from *S* to *R* suggests that in VSO languages, if in fact there are such, *S* can learn to initiate activation of the *T/R* dipole at *R*.) Thereafter, the *T/R* dipole oscillates in phase with the foot dipole of chapter 9. As was mentioned in the discussion of figure 10.6, *T* and *R* need not rebound on every foot.

The first nominal concept to be activated, $N_1$, is the topic, *man*. All substantives can be phonologically realized as either a phonological form Φ or *Pro*. For simplicity, figure 10.8 only diagrams Φ and cerebellar deperseveration for the instance of *man* (i.e., /mæn/). At $t_2$, /mæn/ is output and deperseverated. Now the relative clause $S_{rel}$ is activated. This activation is displayed with an STM arrow because relative clauses are not always attached to nominals. (The ordering of relative clauses, however, is language-dependent and must be learned at LTM traces, which, for simplicity, are not diagrammed in figure 10.8.)

The relative clause, $S_{rel}$, (re)activates the sentential rhythm dipole at *T*. In this case, the topic of the embedded relative clause is also the nominal concept *man*. Since Φ has been deperseverated, *Pro* now becomes active, and *who* is output at $t_3$. At $t_4$, the dipole switches back to *R*. *Aux* and *V* are activated and *is dancing* is output.

Bottom-up deperseveration and rebounds now deactivate *Vp*, $S_{rel}$, and $N_1$. The top-level dipole rebounds to *R*, and the top-level *Vp* is activated. *Aux*, how-

160 • HOW THE BRAIN EVOLVED LANGUAGE

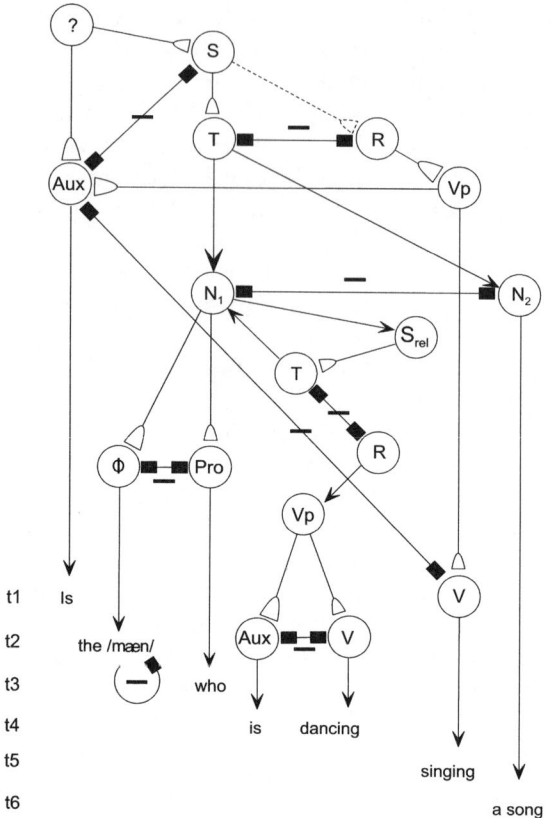

**Figure 10.8.** Generation of questions and relative clauses.

ever, has already been performed and deactivated, so $V = $ *singing* is output at $t_5$. Finally, the top-level $T/R$ dipole rebounds back to $T$; $N_2$ is activated and *a song* is output at $t_6$.

Having dispensed with the need for generative linguistics in this chapter, I should close by crediting generative theory with anticipating many of the key elements of adaptive grammar. Figure 10.8, for example, builds on generative trees, which were generally correct in their structure, if not in their operation. Generative linguistics also correctly predicted the existence of an "abstract, autonomous" grammar, a relational system which functions quite independently of "real-world, substantive" cognition. However, the generative assumption that sentences are generated by movement proved a bad choice of metaphor. Nothing moves. Language needs relevance, and syntax is ordered by topicality. To be useful for survival, grammar must relate to a topic; otherwise, it has no meaning.

• E L E V E N •

# Truth and Consequences

> Consider what effects, that might conceivably have practical bearings, we conceive the object of our conception to have. Then, our conception of these effects is the whole of our conception of the object.
>
> C. S. Peirce, the Pragmatic Maxim
> from "How To Make Our Ideas Clear (1878)

In the last chapter we saw that the topic—that which we are talking about—plays a privileged role in ordering our unfolding motor and language plans, our sentences. But some topics never seem to arise. For example:

> The King of France is bald. (11.1)
>
> You think therefore I am. (11.2)
>
> The human race has never existed. (11.3)
>
> Every bachelor is an unmarried man. (11.4)

One would be very surprised to stray into a discussion on one of these topics at a cocktail party. As we first noted in connection with 11.1, the problem seems to be not so much that such sentences are *false* as that they are simply *void*. They are *meaningless*. Even 11.4, which is very, very true, is very, very trite.

While it is easy to say that sentences like 11.1–11.4 are meaningless and that topics must be meaningful, it is quite a bit more difficult to clarify just what makes an idea meaningful, as Peirce's above attempt illustrates.[1] So let us first try to clarify Peirce. Consider the following sentences:

> Hands up or I'll shoot! (11.5)
>
> Global thermonuclear war will begin any minute. (11.6)

Unlike sentences 11.1–11.4, these sentences bear on matters of life and death. Presumably, they have a great deal of what Peirce would call "practical bearing." Being in the "future tense," neither one would be strictly True before the fact, but either would, in sincere context, be Very Meaningful. Truth and Meaning are not necessarily the same thing.

Sentences 11.5–11.6 are over-the-top, "Hollywood" examples of Meaning, and as a philosopher of science, Peirce would no doubt have found them crass. Only in a footnote to a later (1893) edition of his essay did Peirce deign to give popular expression to his "Pragmatic Maxim":

> Before we undertake to apply this rule, let us reflect a little upon what it implies. It has been said to be a skeptical and materialistic principle. But it is only an application of the sole principle of logic recommended by Jesus: "Ye may know them by their fruits," and it is very intimately related with the ideas of the Gospel. We must certainly guard ourselves against understanding this rule in too individualistic a sense. (Peirce quoted in Wiener 1958, 181n)

The too-individualistic sense against which Peirce warns us was William James's sense of pragmatism. Born the first son of a wannabe Harvard professor (Henry James, the elder), William James succeeded where his father had not. In that previous heyday of American capitalism at the turn of the last century, James popularized Peirce's notion of *pragmatism* with movie marquee rhetoric: "the cash-value of true theories," "truth is what works." Blessed with this clear (some would say pandering) style, James succeeded in becoming a Harvard professor and celebrated as the "Father of American Pragmatism."

By contrast, Peirce was the precocious son of a respected Harvard mathematics professor. He no longer aspired to status. In his 1859 Harvard class book he inscribed the following:

> 1855   Graduated at Dixwell's and entered College.
> Read Schiller's Aesthetic Letters & began the study of Kant.
> 1856   Sophomore: Gave up the idea of being a fast man and undertook the pursuit of pleasure.
> 1857   Junior: Gave up the pursuit of pleasure and undertook to enjoy life.
> 1858   Senior: Gave up enjoying life and exclaimed "Vanity of vanities!"

Disdainful of vanity, Peirce was an intensely original thinker whose writing seems always contorted to avoid the popular clichés of his day. No member of the Get-along-Gang, Peirce was dismissed as arrogant and was little appreciated in his own time. For many years, history regarded Peirce's students and colleagues (including John Dewey, E. L. Thorndike, and his sometimes-antagonist Josiah Royce) more highly than Peirce himself. Had it not been for the patronage of the powerful and influential James, it is possible that Peirce's work would have been totally lost. But as it happened, James's patronage was also patronizing, and his popularization of Peirce's pragmatism with overly simplistic formulae like "the true is the expedient" and "faith in a fact helps create the fact" would have been plagiarism had it been more astute.

In Peirce's view, James confused Truth and Meaning. Meaning resides in the practical consequences of the objects of our conception, but what we find meaningful may not be True. We are fallible. This insistence on "fallibility" led Peirce to rename his philosophy "*pragmaticism*, which [is a term] ugly enough to be safe from kidnappers" (Peirce 1905). As it happened, the times found James's "truth pays" more appealing than Peirce's Jesus. "Truth pays" had more "cash value." Despite James's patronage, Peirce died a failure by Hollywood standards, impoverished and forgotten.

To be fair, we should note that from a psychologist's perspective James' jingles were perhaps defensible definitions of workaday truth, of the rationalizations and convenient fictions of everyday psychopathy. The difference between Truth and Meaning may be less of quality than it is of quantity. I suspect Peirce would not have objected so strongly if James had said, "What works for a long time is true." James was a psychologist of his day, but Peirce was a scientist, and in the scientific ideal, eternal truths work eternally. The problem is that even in science revolutions occur. An Einstein detects a small wrinkle in space-time, and suddenly the entire edifice of Newtonian mechanics is reduced to a convenient fiction of workaday physics. Science's quest for long-term replicability is certainly noble, but for the individual (and sometimes for the species), survival often comes down to short-term, lower case, Jamesian truths. If we can't have truth, we must settle for meaning.

## Truth and Survival in Science

In his classic study of scientific revolutions, Kuhn's central example was the Copernican Revolution (Kuhn 1957, 1962). He paints a picture of licensed Ptolemaic astronomers doodling with epicycles, while outside the ivied halls of the scientific establishment, Copernicus was meticulously noting small discrepancies in measurements and creating the future science of the cosmos. Kuhn examines the historical and sociological dynamics of these paradigm shifts in engaging detail, but for my money, he doesn't sufficiently credit economics. The "cash value" of Copernicus's theory wasn't in its Truth but in its Meaning.

In the fifteenth century, the expansion of maritime trade led intrepid sailors to challenge the popular notion of a flat Earth. Fifty years before Copernicus's text was published in 1543, Columbus had already reached the East by sailing West, and twenty years before Copernicus, Magellan had already circumnavigated the globe (1522). To be sure, Ptolemy thought the Earth was spherical, and the heliocentric system did not directly improve navigation, but it was still the prospect of riches from world trade and the accompanying need for improved navigation by the stars that paid the salaries of Ptolemaic and Copernican astronomers alike. Columbus and Magellan were the ones who conducted the empirical experiments with practical consequences. To paraphrase James, the meaningful theory was what people would buy. By 1543, no one was buying Ptolemy, so

Copernicus could publish *De revolutionibus orbium coelestium*, claiming what experience had found meaningful to also be True. This is what got Galileo into trouble with the Church.[2] The Earth could be round and go around all it wanted, and the Church didn't really care how much money merchants made thereby; it only cared that the heliocentric universe not be declared an Eternal Truth.

Cash value and Truth have been confused in linguistics, too. For Plato and Aristotle, linguistics may have been basic research into eternal truths, but for the Holy Roman Empire, linguistics had practical consequences. It meant language teaching and language learning: teaching and learning the Greek of Scripture and the Latin of the Church. Grammar was a core course of the medieval trivium, and linguists were primarily language teachers . . . at least until the Reformation.

The Reformation was as much a linguistic revolution as it was a social, political, and religious revolution. Luther's original Ninety-Five Theses (1517) are now largely forgotten, but his translation of the New Testament from Latin to German (1534) remains a cultural bible.[3] Coupled with Gutenberg's invention of the printing press (ca. 1456), the mass-produced Lutheran Bible soon had God speaking directly to the people—in German. Job prospects became bleak for Latin and Greek teachers in Germany.

Although German had a Bible, it still lacked the cultural history and prestige the Romance languages had inherited from Latin. But after Jones's theory of evolution (chapter 2), a new generation of linguists set to work reconstructing an earlier Germanic language, a sister to Latin, Greek, and Sanskrit. After Napoleon's demise, this newly discovered classical pedigree became German nobility's title to empire, and while demand may have dwindled for Latin and Romance-language teachers, the aspiring young German philologist could hope for a court appointment to study Germanic and "Aryan." One such aspiring young philologist was Jakob Grimm. In 1808, Grimm was appointed personal librarian to the king of Westphalia. Germanic, unlike Latin and Greek, had left no written literature from which it could be reconstructed, so Grimm and his younger brother, Wilhelm, studied Germanic *oral* literature. In 1812, they published their first collection of fairy tales, *Kinder- und Hausmärchen* (Children's and Home Tales). In 1830, Jakob and Wilhelm Grimm were given royal appointments to the University of Göttingen. Germany was no longer a third-world country, and the Brothers Grimm were no longer publishing fairy tales. By 1835, they had published *Die deutsche Heldensage* and *Deutsche Mythologie* (German Hero Sagas and German Mythology).

At the same time that philology was being celebrated in Germany, linguists were still being employed as language teachers in the United States. Needing a steady influx of immigrants to settle the frontier and expand labor-intensive industry, the young nation founded "grammar schools" which employed linguists to teach English as a second language (ESL)[4] in a New World trivium of readin', writin', and 'rithmetic. In the United States, bilingualism had practical bearings, and language teaching was meaningful. It remained meaningful until World War I limited immigration and the rise of communism discredited bilingualism. To please their patrons and prove their patriotism, Americans became monolingual, and soon language teacher–linguists were no longer needed in the New World either.

After World War II and Hitler's appropriation of the term "Aryan," the job market for philologists collapsed. But as the world's only surviving economy, the United States suddenly found itself an international power. United States soldiers returning home from the war reported with surprise, "No one in Europe speaks English!" Within a decade, study of modern foreign languages became required in every U.S. college and high school. At the same time, the "baby boom" produced a 40% increase in the U.S. birthrate. Eventually, the baby boom became a student boom, and the demand for linguists to teach foreign languages redoubled. Suddenly, linguists could get jobs again.

Leadership in this new, foreign-language teaching movement came from linguists trained in the incompatible methods of philology (the comparative method and the contrastive analysis hypothesis) and psychology (habit formation and interference). As crude as those methods seem today, I still remember my first pattern practice drill in German:

| Willi | Was gibt es denn zum Mittagessen. |
| Hans | Wahrscheinlich Bratwurst. |
| Willi | Ich habe Bratwurst nicht gern. |

But when the first cohort of multilingual U.S. students and I went abroad, eager to strike up conversations about bratwurst, we found that everybody else in the world had already learned English!

Almost simultaneously, oral contraceptives were invented and the baby boom became a baby bust. Within a generation, English became the lingua franca of the "new world order." In the United States, there was suddenly no longer a pressing national need for foreign languages. Before long, colleges and universities had removed their foreign-language requirements. Soon there were few foreign-language students, and there were fewer jobs for foreign-language teachers.[5] Fortunately, there were other job opportunities for American linguists, but they were top secret.

At the heart of German war communications in World War II was the *Enigma Machine*. The Enigma Machine was a kind of cryptographic cash register which took in a message, letter by letter, and then, by a complex system of gears, put out an elaborately transformed and encrypted code. For example, if today were Tuesday and *e* were input as the 1037th letter of the message, then *x* might be the output code. To defeat Germany, the Allies needed to defeat the Enigma Machine, and they needed to do it *fast*. As it happened, in 1936 Alan Turing published a paper which mathematically described a universal cryptographic cash register, one which could be configured to emulate any kind of real cryptographic device. With the outbreak of hostilities, the cash value of Turing's theory skyrocketed. The German's Enigma Machine was a "black box": from enemy actions, cryptographers could see what had gone in, and from intercepted enemy radio messages they could see what had come out, but they couldn't see how it did it. The black box had to be "reverse-engineered." To that end, the Allies immediately began a major war program to build a "Turing machine" which could emulate the German's Enigma Machine. At the end of the war, the Turing machine was upstaged by the atomic bomb, but the generals knew that the triumph of the Allies was in large measure the triumph of the Turing machine and of a new linguistics, a *computational* linguistics.

In 1949, on behalf of the U.S. military and espionage establishments, Warren Weaver of the Rand Corporation circulated a memorandum entitled "Translation" proposing that the same military-academic complex which had broken the Enigma code redirect its efforts to breaking the code of the Evil Empire, the Russian language itself. Machine translation became a heavily funded research project of both the National Science Foundation and the military, with major dollar outlays going to the University of Pennsylvania and the Massachusetts Institute of Technology. In 1952, Weaver outlined a strategy before a conference of these new code-breakers. The strategy was to first analyze, or *parse*, Russian into a hypothetical, abstract, universal language, which Weaver called *machinese*, and then to *generate* English from this *machinese*. At MIT, the machine translation effort became organized under the leadership of Yehoshua Bar-Hillel, and in 1955, Bar-Hillel hired a University of Pennsylvania graduate student who just happened to have written a dissertation outlining a theory for generating English from machinese. His name was Noam Chomsky, he called machinese "deep structure," and his theory was "generative grammar."

By 1965, however, Bar-Hillel had despaired of achieving useful machine translation. The main problem was that the MIT Russian parsing team had "hit a semantic wall." It never succeeded in producing deep structures from which Chomsky's theory could generate English. In describing this semantic impasse, Bar-Hillel noted how hard it would be for a machine to translate even a simple sentence like

Drop the pen in the box. (11.7)

The problem was meaning. The problem with 11.7 was that *drop*, *pen*, and *box* all have three or more *senses* (meanings with a small *m*). Theoretically, some $3^3$ different sentences could be generated from a deep structure containing just those three substantive terms. Consider for example *11.8 and 11.9:

*Drop the pen in $the_{det}$ $box_{verb}$. (11.8)

?Drop the $pen_{playpen}/pen_{ballpoint}$ in the $box_{trailer}/box_{container}$. (11.9)

Sentence *11.8 is fairly simple to solve. It can be rejected as an ungrammatical sentence by a simple generative grammar rule, something like *a verb may not immediately follow a determiner.* But 11.9 is more problematic. In 11.9, both *pen* and *box* are nouns. Each could be translated by two different Russian words, but how was a poor computer to know which one was the right one? Ostensibly for this reason the U.S. government gave up on machine translation in 1966

(ALPAC 1966). In point of fact though, the reason was more economic. As evil as the Evil Empire might have been, the United States was at the time gearing up its war on Vietnam, and Russian-English machine translation was not going to be of immediate help. Some research had to be sacrificed for the war effort. Physics, computer science, and mathematics all took one step back, and linguistics was volunteered.

Two years later, in 1968, three discoveries obviated the ALPAC report's criticism of machine translation: (1) Bobrow and Fraser's description of the augmented transition network, (2) Fillmore's case grammar, and (3) Quillian's semantic networks.

In fact, Bar-Hillel was much too skeptical. Chomsky (1965) had already made considerable progress on sentences like 11.9 with his work on *selectional restrictions*. We can drop a $pen_{ballpoint}$ into a $box_{container}$, but we can't drop a $pen_{prison}$ into a $box_{container}$. If our friendly neighborhood lexicographer were to define $pen_{ballpoint}$ to have the feature *+object* and $pen_{prison}$ to have the "semantic feature" *+institution*, then a simple grammar rule restricting *drop* to the selection of a direct object which was either *+object* or *–institution* would reject *11.10:

*Drop the $pen_{prison}$ in the box. (11.10)

This is a kind of agreement rule. We could say the semantic features of the verb must agree with the semantic features of the direct object, but this solution still posed several technical problems. The first problem was finding a way to compute agreement between separated phrases—so-called *long-distance dependencies*. Within just two years of the ALPAC report, Bobrow and Fraser (1969) solved the general problem of long-distance dependencies with the augmented transition network (ATN). Whereas the lambda calculus and the pushdown-store automaton had two Turing machines working together, the ATN formalism had *three*: one for program, one for data, and one for agreement.

A second problem arose when the verb and the direct object underwent a passive transformation, as in 11.11:

The pen was dropped in the box. (11.11)

In this case, semantic agreement needs to be enforced between subject and verb, not between verb and direct object. This problem was also solved in 1968 by Fillmore's *case grammar*, which as we saw in chapter 10 replaced terms like "subject" and "direct object" with terms more appropriate to computing semantic agreement on selectional restrictions, terms like "actor" and "patient."

Finally in 1968, Quillian published his ideas on *semantic networks*. The gist of Quillian's idea is illustrated by figure 11.1. *Pen* has (at least) three senses. $Pen_1$ *is a* tool. This is represented in figure 11.1 with an *ISA* link from $pen_1$ to *tool*. In figure 11.1, a *tool* also *ISA instrument*, which illustrates the easy linking of a semantic network to case grammar. $Pen_2$ *ISA* enclosure, as is $pen_3$. $Pen_1$, however, is *FOR writing*, while $pen_2$ is *FOR animals* and $pen_3$ is *FOR criminals*.

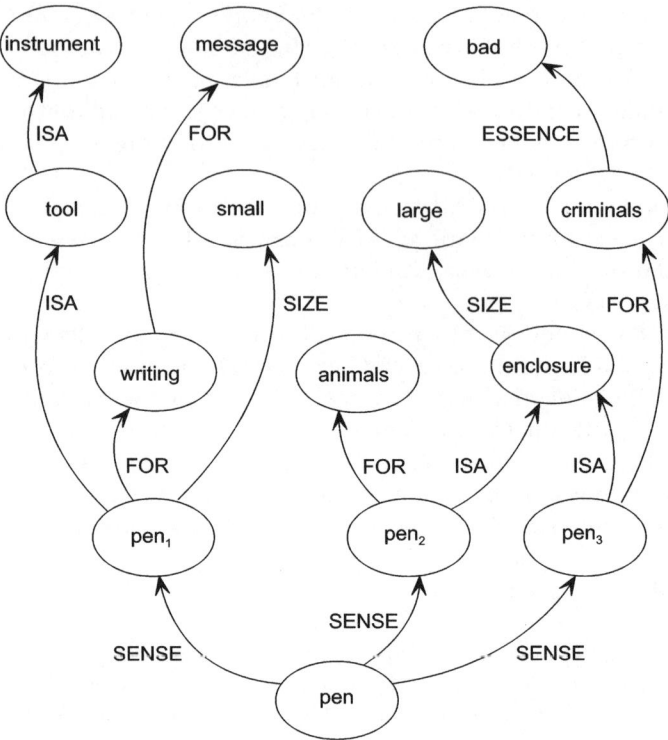

**Figure 11.1.** Fragment of a semantic network.

Quillian's one representation scheme greatly simplified the organization of a computational lexicon. It could encompass verbs, nouns, and other parts of speech. Moreover, the semantic network was also a compact representation. Instead of repetitively coding tens, and possibly hundreds, of features on each word in the lexicon, features could be inherited along ISA links. Thus, each meaning element needed to be represented only once in the net. For example, having the links *dog ISA mammal* and *mammal ISA animal*, we don't need to represent the fact that a *dog* can follow links from *mammal* to the further inference that a *dog ISA animal*.

In psychology, Quillian's work was closely followed by a series of "spreading activation models" of word association. Morton (1969, 1979) proposed his "logogen model" as a psychological account of word recognition during reading. In this model, concept nodes (logogens) fired when they detected a word, and the more they fired, the more easily they would fire the next time. In this manner, they learned the word they detected. Contextual activation was also seen as spreading between logogens along semantic links. Such spreading activation models featured a Hebbian explanation of why frequent words are recognized more quickly than infrequent words, and they also offered an explanation of certain "priming effects." For example, subjects recognize the word *nurse* more quickly after having first been "primed" by hearing or seeing the word *doctor* (Swinney 1982; also see Small et al. 1988 for a review and related papers).

## Perceptrons

Ever since Ramón y Cajal, it had been obvious that the brain is a massively parallel processor, but no one knew what this *meant*. There had been a few early attempts to model parallel computation (e.g., McCulloch and Pitts 1943), but it wasn't until the late 1950s that researchers began to think seriously of the serial, digital computer as a starting point for parallel models of the brain. In 1958, Rosenblatt popularized the most famous early parallel-computing model, the *perceptron* (Rosenblatt 1958). For a brief time, parallel computer technology competed with serial computer technology, and as early as 1965, the Illiac IV, a 64,000-element parallel computer, became operational.

It seems, however, that no one could figure out how to program the Illiac IV, and in 1969, Marvin Minsky and Seymour Papert of MIT published *Perceptrons*, certifying what experience had found meaningful to also be True. Finding that perceptrons could not even compute the elemental and essential Boolean operation of XOR, Minsky and Papert concluded that "there is little chance of much good coming from giving a high-order problem to a quasi-universal perceptron whose partial functions have not been chosen with any particular task in mind" (1969, 34). As it happened, only three years later a celebrated young associate professor in Papert's laboratory published the first of several papers which showed not only that the brain was a parallel processor but also that it computed XORs ubiquitously (Grossberg 1972b, 1972c).

But no one was buying parallel computing any longer. After Minsky and Papert's critique, funding for parallel computation research dried up, and further progress in the field was not widely recognized until 1982 when Hopfield published his theory of "content-addressable" memory. Hopfield's memory was reminiscent of earlier work in perceptrons, but in the intervening fifteen years *microprocessors* had been developed. IBM had just introduced its personal computer, and it was now easy to imagine bringing massively parallel computers to market.

Other researchers, following these models, extended the basic, single-layer perceptron into multilayer perceptrons. In an unpublished doctoral dissertation, Hinton (1977) had described such a system, and in the new climate, Ackley, Hinton, and Sejnowski (1985) elaborated that system as an extension of the Hopfield model. By sandwiching a layer of Hopfield-like neurons between layers of input and output neurons, the resulting multilayered perceptron, or *Boltzmann machine*, could not only remember but also actively classify. At nearly the same time, a number of researchers, most prominently Rumelhart, Hinton, and Williams (1986), added to this architecture an error correction technique known as *back-propagation*. Back-propagation models employ a fundamentally cerebellar design and so are not particularly useful for modeling higher cognitive functions like language (Loritz 1991). Nevertheless, they are relatively easy to implement on serial computers. As a result, they have found a ready resonance among computer scientists, and they furnished psychologists and linguists (e.g., Goldsmith 1993, 1994) with new metaphors of mind and language. In the sandwiched, or "hidden," layers of multilevel, back-propagating perceptrons, the convergence of inputs and divergence of outputs are so complex that simple and direct interpretation of their activation patterns becomes impossible: the associations between input and output are effectively *hidden* from the researcher.

Which brings us, by a commodius vicus of recirculation, back to meaning. It is much the same in real thought. Between the input and output layers of language, the intricate connections of our meaning are hidden, shrouded in complexity. Even if I could communicate, instant by instant, every synapse of meaning, every past and possible $x_i$ and $z_{ij}$ in *my* brain, my meaning would still be *hidden* in the $10^{7,111,111}$ different ways *you* could interpret those weights. Of course, when our tangled nets of words and concepts happen to reach a place of relative understanding, we can give that place a name. And when we look back on the places we have been and the path we have followed, we can call it logic. If we keep our names simple enough, if we only seek to agree on what is 1 and what is 0, then we might even find that this path of logic leads to a kind of Truth. But Truth does not guarantee survival, and this linear logic is not Meaning; it is only a trail of linguistic cairns.

This brief history of linguistic science gives us pause to reflect on how thought is like meaning. In the preceding chapters of this book, I have described thought as a social state of neurons, neurons involved in acts of communication. In this chapter, I have described meaningful linguistic theories as a social state of *Homo loquens*. Thought and meaning are both only stable when neurons or people are in a state of communicative resonance. This standing wave of communication may appear to be an eternal, true state, especially to those individuals who are locked in its resonance, but at the periphery, the environment is always changing, and there are other inputs which other populations are detecting and encoding. All it takes is an unexpected turn of events—a burst of nonspecific arousal, a Columbus sailing west and arriving east—for a revolution to occur.

We shouldn't be surprised to find these similarities between thought and meaning, between neural society and human society. After all, a human being is just a neuron's way of making another neuron. This may be self-similarity pushed too far (I am reminded of Daniel Dennett's "a scholar is a library's way of making another library"), but this is the direction in which self-similarity leads. Even it doesn't form a useful basis for either neurosurgery or public policy, it does often supply a useful metaphor, one we can use to develop meaning, if not truth.

I am often asked by my students, "Is it true we use only 10% of our brains?" To which I respond, "Does a society use only 10% of its people?" Of course, some do, but in the main, most societies use almost all of their people almost all of the time. The real question is *how* does it use them? Lateral competition exists in sibling rivalry and office politics as surely as it exists in the cerebrum, and the human organism swims ahead in history, oscillating its tale from left to right, unsure of where it's going but trying always to survive.

• T W E L V E •

# What If Language Is Learned by Brain Cells?

The generative deduction held that children learn their first language effortlessly. As the theory went, an innate language acquisition device (LAD) stops operating somewhere between the age of six and twenty. Children, thanks to their LAD, were supposed to "acquire" language perfectly and "naturally" (Krashen 1982). Adults, however, were left with no alternative but to "learn" second languages in an "unnatural" and suboptimal manner. Thus, generative theory was forced to posit *two* mechanisms, which it could not explain, much less relate: one to account for the facts of child language "acquisition" and another to account for the facts of adult language "learning."

Adaptive grammar explains both child and adult language learning with one set of principles which evolves similarly in both ontogeny and phylogeny and operates similarly in both brain and society. After all, it is not clear by what yardstick generative philosophers measured child labor. We adults say we find second languages harder to learn as our years advance, but it may just be that we value our time more as we have less of it. By simple chronological measure, normal children do not learn language all that quickly. It takes children six years or so to learn the basic sounds, vocabulary, and sentence patterns of their mother tongue. By some accounts, this process extends even beyond the age of ten (C. Chomsky 1969; Menyuk 1977). Adults can learn all of this much faster than children (Tolstoy, it is said, learned Greek at the age of eighty), even if there is some debate as to how well they learn it.

There is a widespread consensus that adults never learn to *pronounce* a second language well—spies and opera singers notwithstanding—but few would argue that fastidious pronunciation is a cognitive ability of the first order. Adaptive grammar assigns fastidious pronunciation to the cerebellum, and just as few violinists who take up the instrument in retirement achieve the technical capacity of a five-year-old prodigy, the adult who attempts to emulate Tolstoy may find his cerebellum regrettably nonplastic: as we have seen, the cerebellum lacks the capacity to learn new patterns through rebounds. In adaptive grammar's analysis, the cerebellum is as essential to language as rhythm, and in this chapter we will even relate it to several language disorders, much as Kent et al. (1979) related it to plosive ataxia in chapter 8. But pound for pound, the cat has a bigger cerebellum than *Homo loquens*, so the cerebellum seems an unlikely site for any uniquely human language acquisition device.

On the other hand, the cerebrum is never so plastic as at the moment of birth, when it is—sans sight, sans smell, sans taste, sans dreams, as Changeux (1985) puts it—as rasa a tabula as Locke could have imagined. Its ability to adapt thought and language to an unstable environment has secured *Homo loquens* a singular history of survival. Not even T. Rex ruled his world with such power over life and death. In the preceding chapters, we have tried to follow that history through evolutionary time, through the phylogeny of *Homo loquens*. We will now recapitulate those chapters in an examination of the ontogeny of language in the individual human.

## Prenatal Language Learning

There is no disputing the fact that language is innate. Without genes, rocks would be the highest life form. But there are only some $10^5$ genes in the entire human genome, and only some $10^3$ of these are uniquely human. Even if all of these were genes for language, and even if they are allowed to operate in combinations, there simply aren't enough uniquely human genes to specify uniquely human language. For language, we must still recruit the genes of our phylogenetic ancestors, and over this still-sparse genetic scaffolding, language must still be learned. This learning begins even in the womb, as the environment impinges upon brain development.

The environment of the womb is not all so hostile as the primordial soup. Almost from the moment of conception, the mother's heartbeat envelops the child, and its first neurons learn the rhythm of humanity. This is not quite yet the rhythm of language, to be sure, but soon the child's own heart begins to beat, and the first rhythmic foundations are laid for language and all serial behavior. The human child is conceived, and the fertilized egg begins to divide into multiple cells, forming first a blastula and then a two-layered gastrula. Within two weeks, the vertebrate notochord begins to clearly differentiate itself from the rest of the gastrula. By four weeks, the cerebrum has differentiated itself from the notochord and attained about a size and shape proportionate to an Ordovician fish. At the same time, an otocyst forms from the ectodermal membrane. This otocyst will become the inner ear.

By seven weeks, the fetal brain has achieved roughly the cerebral development of a reptile. Within another week, the cochlea will be completely formed and coiled. By twelve weeks, the cerebrum will have achieved rough proportionality to the brain of an adult cat, and by about four or five months, the child's full complement of neurons will have been produced by mitotic division.

## Critical Periods

In the wake of Hubel and Wiesel's work suggesting a *critical period* for the organization of visual cortex, Lenneberg (1967) suggested that a similar critical period might exist for language acquisition. It was originally widely believed that critical periods were, like growth and language development, essentially irreversible. Just as there was assumed to be a detailed genetic plan for the development of notochord and otocyst, there was assumed to be a genetic plan for the organization of the brain, down to the detail of Hubel and Wiesel's ocular dominance columns. Lenneberg hypothesized that the organization of language cortex into yet-to-be-discovered grammar-specific structures resulted from a similar genetic plan. Generative philosophy held that language is not *learned* so much as it *grows* in the child.

As it turns out, however, the Hubel-Wiesel critical period is reversible. Kasamatsu et al. (1979) poisoned the noradrenergic arousal system of young cats and observed that this diminished plasticity in accordance with the Hubel-Wiesel critical period. But then they added noradrenaline to the cortex of these adult cats, after their critical period had supposedly ended, and they found that this *restored* plasticity. Coupled with the effects cited in chapter 3, where I noted that noradrenaline can increase the signal-to-noise response ratio of neurons, adaptive grammar interprets Kasamatsu's nonspecific noradrenalin suffusion as having effected a kind of nonspecific arousal (NSA) and an ensuing plastic rebound. Such a reversal of plasticity would be a most unusual property for a growthlike *genetic* process. But as we have seen, no detailed genetic plan is necessary for ocular dominance columns and tonotopic maps to develop. They can *self-organize* with no more detailed a plan than that of an on-center off-surround anatomy.

## Lateralization

Broca's discovery that language is normally lateralized to the left cerebral hemisphere has attracted curiosity and speculation ever since its publication, and the fact of lateralization figured prominently in Lenneberg's speculations. But no sooner had Lenneberg proposed that lateralization occurred during childhood than Geschwind and Levitsky identified a lateralized cerebral asymmetry in the superior planum temporale of human and animal fetuses (Geschwind and Levitsky 1968; Geschwind 1972; Geschwind and Galaburda 1987). This area of the left hemisphere between Heschl's gyrus and Wernicke's area was found to be larger than corresponding regions of the right hemisphere in 65% of the brains studied (figure 12.1). The left and right areas were found to be equal in 24% of the cases, and a reverse asymmetry was found in 11%. Noting that this enlarged planum temporale was less reliably found in boys, who are also more prone to be left-handed, language-delayed, and dyslexic, Geschwind hypothesized (1) that abnormal symmetry (non-lateralization) could be caused

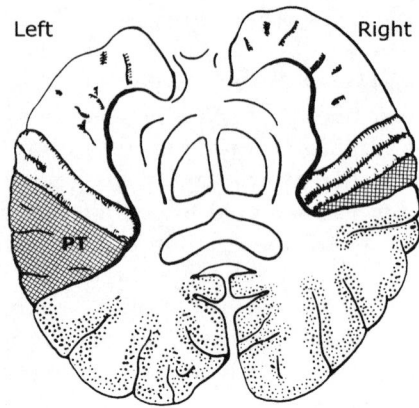

**Figure 12.1.** Cerebral asymmetry in the superior planum temporale (PT). (Geschwind and Levitsky 1968. Reprinted by permission of the American Association for the Advancement of Science.)

by the same effusion of testosterone that, while effecting sexual development in utero, also affected brain development, and (2) that many, if not most, left-handers might have been brain-damaged in infancy in this way or another.

At first it was thought that such physiological lateralization of language into the left hemisphere was a critical and uniquely linguistic process. But further research discovered that birds and bats and all manner of beasts exhibit brain asymmetries, and lateralization is no longer widely thought to be a language-specific process. Rather, adaptive grammar suggests that all serial behaviors are normally lateralized to one hemisphere because, at their lowest levels, serial long-term memory (LTM) primacy gradients optimally form within a barrel's inhibitory surround. This would encourage (but not strictly require) atomic serial patterns like syllables and words to organize in compact, intrahemispheric regions like Broca's area, Wernicke's area, and the nominal and verbal regions of temporal cortex described by Damasio et al. (1996). If *Homo loquens* needed to locate a few extra serial LTM gradients anywhere, the larger superior planum temporale, which surrounds the left koniocortex, and the planning regions of *Homo loquens*' extensive frontal cortex in and anterior to Broca's area would be good places to put them.

Kim et al. (1997) have found evidence supporting this explanation. They conducted an fMRI study of six fluent early bilinguals who learned their second language (L2) in infancy and six fluent late bilinguals who learned their L2 in early adulthood. MRI scans of the late bilingual subjects speaking both L1 and L2 (which varied widely) revealed significantly distinct regions of activation, separated by a mean of 7.9 mm, for each language within Broca's area. Late bilinguals' L1 morphophonemic motor maps had developed in close proximity, as if their extent were limited by some factor like inhibitory radius. Their L2s developed in a separate area, as if the L1 area was already fully connected.

Early bilinguals did not exhibit two distinct regions; presumably, their morphophonemic motor maps all developed together in a single area.[1]

Although left lateralization may be a normal event in language development, it does not seem to be a necessary event. In deaf sign languages Heschl's gyrus and the planum temporale are only marginally activated. If anything, sign language seems to be doubly lateralized—with its temporal, planning functions lateralized to the left hemisphere and spatial functions lateralized to the right. In childhood aphasia, we also find the heartening fact that recovery is possible and often total. The plasticity of the brain allows the child aphasic's right hemisphere to completely take over the function of the better-endowed but injured left hemisphere. Left-handers wind up being, if anything, slightly superior to right-handers in their verbal skills. The only necessary loss seems to be a small delay in the (re)learning of language, in the translocation of some atomic primacy gradients to the contralateral hemisphere. If the race is to the quick learner, Geschwind's asymmetry could have conferred a selectional advantage on *Homo loquens* as a species, but the brain is plastic and language is learned, so mother and family only need to protect any individual symmetric boy or left-hander for a few extra months until his plastic brain can adapt.

Environmentally induced critical periods

While lateralization of the planum temporale might appear genetic to us, to your average neuron a flood of testosterone must occur as an environmental cataclysm. In fact, numerous exogenous, environmental effects have recently been found to affect the developing brain, with indirect consequences for language. In 1981, West et al. demonstrated that exposure of rat embryos to ethyl alcohol could disorganize development of the hippocampus. In figure 12.2B, hippocampal pyramidal cells that normally migrate outward to form a white band in the stained preparation of figure 12.2A have clearly failed to migrate properly under exposure to ethyl alcohol. It has always been known that children of alcoholic mothers were at risk for birth defects, and these were often

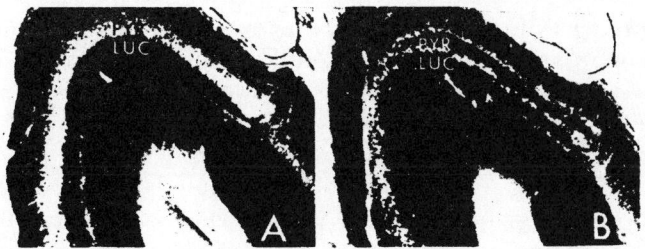

**Figure 12.2.** Ethyl alcohol disorders neuron migration in the hippocampus of fetal rats. Normal pyramidal cells form a coherent band in *A*. In *B*, after exposure to ethanol, the pyramidal cells become physiologically disordered during neurogenesis. (West et al. 1981. Reprinted by permission of the American Association for the Advancement of Science.)

further associated with learning and language disorders, but seeing is believing. In West et al.'s work, it became clear that the developing brain could be *physiologically* disordered. The teratogenic effects of alcohol upon the developing brain have been consistently demonstrated when the alcohol is administered moderately late in gestation. (Earlier poisoning with alcohol can simply kill the fetus.)

Behavioral manifestations of ethanol-induced disorders are often less obvious. In humans, because any linguistic symptoms of fetal alcohol syndrome (FAS) cannot be observed until years after birth, such neurological damage might be far more prevalent than the number of cases diagnosed and reported at birth (currently about 1% of all live births).

Teratogens, disease, and genetic defects

Testosterone and alcohol are not the only "environmental" *teratogens*[2] to threaten the developing fetus. Lead, mercury, common prescription drugs like retinoic acid (Retin-A™ or Accutane™), and common diseases can all cause birth defects and/or learning disorders. Maternal *rubella* (German measles) causes very serious birth defects, including deafness. Even maternal *varicella* (chicken pox) has been linked to Reye's syndrome. Mednick (1994) reported a study of Helsinki children in which maternal influenza during the second trimester resulted in an exceptionally high incidence of subsequent adolescent schizophrenia, and it would not be surprising to find many more-subtle but serious prenatal environmental hazards to language and cognitive development.[3]

Other disorders, for which no environmental cause is apparent, are thought or known to be genetic. Fifty years ago, virtually no learning disorders were known except Down syndrome and a broadly defined "mental retardation." Today it appears there may be hundreds. The following sections will briefly describe a few of these which especially involve language learning.

*Williams Syndrome*

Schizophrenia is a poorly understood and ill-defined disease that often does not manifest itself until adolescence, so it is unclear what specific or indirect impact influenza might have upon neural development or language development, but Williams syndrome may offer an approximate example. Briefly, children with Williams syndrome present IQs in the retarded range but normal grammar and language. One eighteen-year-old spoke of her aims in life as follows: "You are looking at a professional book writer. My books will be filled with drama, action, and excitement. And everyone will want to read them. . . . I am going to write books, page after page, stack after stack. I'm going to start on Monday" (Reilly et al. 1991, 375). Unfortunately, her plan was delusional. At a local, grammatical level of planning and organization, all appears well, but globally, her plans and organization are incoherent in a manner vaguely reminiscent of schizophrenia or even Wernicke's aphasia. Wang and Bellugi (1993) report that, in contrast to Down syndrome, Williams syndrome is characterized by preservation of prefrontal cortex (where adaptive grammar locates grammaticomotor planning) but diminished lateral cortex and interhemispheric callosal fibers. This appears to also be associated with the loss of one copy of a gene that makes elastin, a protein that is the chief constituent of the body's elastic fibers (Sadler et al. 1994). It would seem that if interhemispheric connections are retarded, development of the arcuate fasciculus might also be retarded in Williams syndrome, perhaps sometimes causing symptoms of conduction aphasia to also be present.

*Tourette Syndrome*

Whereas Williams syndrome is characterized by loquacious nonsense, Tourette syndrome is marked by motor tics and/or terse, involuntary, and often vulgar interjections. Recent research ties Tourette syndrome to a hyperactive dopaminergic system involving the caudate nucleus (Wolf et al. 1996). Insofar as Tourette syndrome might be construed as a kind of hypertonic disorder related to obsessive-compulsive disorders (it is exacerbated by dopamimetic drugs like L-dopa, which are used to treat Parkinson's disease), it calls to mind adaptive grammar's earlier suggestion that a tonic articulatory posture might account for Klein et al.'s (1994) finding of selective L2 activation in the putamen. Tourette syndrome is genetic, but it is *incompletely penetrant*: identical twins develop the disorder to differing degrees, suggesting that nongenetic factors, including prenatal influences, can mitigate or exacerbate Tourette syndrome (Leckman et al. 1997).

*Offbeat Dysphasia*

Most language-learning disorders report different and subtler symptoms than Williams or Tourette syndrome. Gopnik (1990; Gopnik and Crago 1991; Ullman and Gopnik 1994) reported on a family in which sixteen of twenty-two members in three generations had been independently diagnosed as *dysphasic*. (The term "dysphasia" is typically used to describe a milder aphasia, not of traumatic origin.) Unlike in Williams syndrome, the dysphasics' global, semantic processing appeared to be unaffected. Rather, they produced local errors such as 12.1:

*The little girl is play with her doll. (12.1)

With a series of tests like the wug test (chapter 9), Gopnik and Crago observed that these dysphasics "could not process grammatical features." In the popular press, it was widely reported that Gopnik had discovered "the grammar gene."[4]

Cross-linguistic work on agrammatism shows a rather similar pattern (Menn and Obler 1990). In wug tests, Broca's aphasics omit free morphemes like *and* or *of* and bound syllabic forms like *-ing* much more frequently than bound, nonsyllabic morphemes like *-s* and *-ed*. In adaptive grammar, these are the forms which are generated "on the offbeat" out of the relation gradient. This analysis is reminiscent of Leonard's theory that omitted morphemes will be short in duration relative to surrounding units (Leonard 1996),[5] but it further suggests that rhythm disorders may underlie many such linguistic disorders. Rice and Wexler (1996, 1253; see also Lahey et al. 1992, 390) criticize Leonard's account on several points: (1) children are more likely to omit *is* than *-ing* in sentences like *He is going to the store*, (2) children are more likely to produce the contracted *is* in *He's going* than the uncontractible *What is it?* (3) dysphasic children are apt to omit *does* in sentences like *Does he want a cookie?* What appears to unify Leonard's and Rice and Wexler's accounts in these cases is that dysphasic children are more likely to omit unstressed (offbeat) grammatical morphemes.[6]

*Dyslexia*

When we read, our eyes do not scan the line of text in a smooth, uniform sweep. Rather, they jump from fixation point to fixation point in a series of *saccades*, each several words long. Some children have difficulty learning to read and are diagnosed as *dyslexic*. Among dyslexics, there is often a tendency for the eye saccades to overshoot their target (Ciuffreda et al. 1985; Eden et al. 1994). This is reminiscent of the cerebellar overshoots measured in the Kent et al. study (1979) we reviewed at the end of chapter 8, and indeed, the same literature has often linked dyslexia and cerebellar disorders.[7]

In a closely controlled, double-blind study, Kripkee et al. (1982) reported the successful treatment of a small cohort of dyslexics with monosodium glutamate (MSG). As we have seen, glutamate is a ubiquitous excitatory neurotransmitter, so superficially these findings suggest that at least one form of dyslexia might result from an underexcited cerebellum. The explanation is complicated, however, by the fact that orally ingested glutamate does not readily cross the blood-brain barrier.[8] However, several "glutamate holes" do exist in the blood-brain barrier, and one leads to stimulation of the pituitary gland of the hypothalamus. The hypothalamus stimulates the sympathetic nervous system, an adrenergic system that raises heart and respiration rates. The pituitary gland produces about eight major hormones, including ACTH, *adrenocorticotrophic hormone*. ACTH, in turn, stimulates the cortex of the adrenal glands, producing adrenaline on the other side of the brain barrier, but the nervous system carries this information back to the brain, where noradrenaline levels rise in response. As we saw in chapter 3 and in Kasamatsu et al.'s results (1979), noradrenaline appears capable of increasing a cell's signal-to-noise ratio and even reversing plasticity.

Kripkee's clinical observation of one subject followed over many years (personal communication) offers a series of deeper clues. The link to MSG was first noted after this subject had dinner at a Japanese restaurant, but the subject had also noted remission of dyslexic symptoms in times of stress: on one occasion she reported reading *The Amityville Horror* (Anson 1977) cover to cover with no dyslexic difficulties or reading fatigue. Apparently, adrenaline-raising, scary books might be as effective a dyslexia therapy as MSG!

But even more appears to be involved than glutamate and noradrenaline. Ultimately, Kripkee found that MSG ceased to benefit his longitudinal subject at the onset of menopause. Dyslexic symptoms reappeared and were corrected only when the subject began hormone replacement therapy.

*Temporal Processing Deficits*

Tallal (1973, 1975) has long studied populations of "language-learning impaired" (LLI) children whose grammatical comprehension and general language processing are deficient and whose speech discrimination is notably deficient in the perception of voice onset time and formant transitions. Tallal et al. (1996) reported that a cohort from this population showed general and long-term linguistic improvement after four weeks of training on speech whose tempi had been digitally slowed and whose formant transitions had been amplified by 20 dB. One of the effective training techniques was a competitive computer game which rewarded the children with points for correctly processing faster speech segments (Merzenich et al. 1996).

*Attentional Deficit Disorders*

Perhaps the most commonly diagnosed learning disorder is "hyperactivity," or attentional deficit disorder (ADD). Paradoxically, hyperactive children are effectively treated with Ritalin™ and other amphetamine-like drugs. These drugs *raise* noradrenaline levels, and in normal people they do not suppress but rather *cause* hyperactivity. Grossberg (1975, 1980) explained this in terms of a *quenching threshold* (QT). An on-center off-surround field has a QT such that neural populations activated below the QT are suppressed while populations activated above the QT are amplified. This is the same process which results in contrast enhancement and noise suppression, as discussed in chapter 5. Our ability to "concentrate" or "focus attention" on one particular object or event is related to our ability to raise the QT so that all but that object of attention is inhibited. In Grossberg's analysis, "hyperactive" children are actually *under*aroused in the sense that their QT is set too high: nothing exceeds the QT for attentional amplification, so they are easily distracted by objects and events in the environment which they should be able to keep inhibited. Eventually, Grossberg's analysis was accepted, and we no longer say that such children are "hyperactive" but rather that they have an "attentional deficit."

The preceding is but a small sample of the increasing array of learning disorders which twentieth-century science has brought to light. Adaptive grammar doesn't pretend to have all the answers to these disorders, but it does suggest some common themes. For example, as noted in chapter 7, the auditory tract arising from the octopus cells of the cochlear nucleus seems designed to detect fast events like the formant transitions of initial consonants. If these cells do not sum to threshold fast enough, adaptive grammar would predict deficits similar to those of Tallal's LLI subjects. But why and how could competitive computer games improve and have a lasting effect upon temporal linguistic performance?

According to adaptive grammar, Ritalin™, scary books, MSG, and exciting computer games could all mitigate temporal processing deficits by stimulating noradrenergic and/or glutaminergic arousal systems: a faster system should, in general, be able to process faster stimuli. Kasamatsu's findings suggest that increases in noradrenaline could also promote learning and plasticity by operating directly on synaptic membranes, apparently by facilitating NMDA response to glutamate. There may be good reason for professors and brainwashers to favor pop quizzes and high-stress environments![9]

The conditions I have been characterizing as dysphasia are often referred to as *specific language impairments* (SLI). This means that the condition is thought to be specific to language, and the existence of such "specific language impairments" is often used to validate the existence of an autonomous "language module." The term "specific language impairment" may be preferable to "dysphasia" insofar as it forestalls the prejudicial implication that people who are so impaired are significantly impaired in other respects. The preceding example of dyslexia attributable to indelicate eye saccades should serve to illustrate that such an implication is far from justified. However, it also illustrates the danger of assuming that because a disorder specifically affects language, there must exist a language-specific module of mind. The cerebellum exerts fine motor control over many aspects of behavior; but it is by no means "language-specific." It just so happens that in modern life there is no other behavior so fine, so common, and so important as eye saccades during reading. Thus, a cerebellar dyslexia may appear to be language-specific, but the cerebellum itself is hardly a language-specific "module." Language is so complex, and it depends so heavily upon the adaptive elaboration and interplay of so many different brain subsystems, that talk of a monolithic, autonomous language module no longer serves as a particularly useful metaphor.

Morphology

From chapter 8 onward, rhythmicity and morphology have emerged as central players in the evolving organization of language. Morphology has been closely examined in the first-language-learning literature since Brown 1973, but it has recently come under increasing attention. In part, this interest has been driven by connectionist models of learning English regular and irregular past tense forms (Rumelhart and McClelland 1986a; MacWhinney and Leinbach 1991; Plunkett 1995; see Pinker and Prince 1988, 1994, for critiques), and in part it has been driven by studies of dysphasia and specific language impairment.

Whereas Tallal's LLI subjects were impaired at time scales of 0–100 ms, the morphological disorders of Gopnik's dysphasics involved grammatical morphemes like *-ing*, which may have durations of 200 ms and more. In chapter 9,

I posited two rhythm generators, one for syllables, on the former time scale, and one for metrical feet, on the latter time scale. These rhythm generators might be differentially involved in the two disorders. However, Gopnik's subjects also had significant phonological impairments, and Tallal's subjects may also have had grammatical impairments like those of Gopnik's subjects, so it is possible that small temporal deficits in motor and sensory processing like Kripkee's and Tallal's can become compounded to also produce dysphasic deficits. Even dyslexia is frequently linked to subtle morphological and phonological deficits. Eden et al. (1994) noted an inverse correlation between dyslexia and the ability to process pig latin—which the reader will now recognize as a metathetic skill.

A second theme underlying the preceding results is that of rhythm and cerebellar function. Until recently, the cerebellum has been widely discounted as a player on the linguistic and cognitive stage. The cerebellum's effects are often subtle (Holmes [1939] reported that people can compensate adequately for even severe cerebellar damage), and its obvious role in motor behavior fixed its theoretical position on the machine side of the Cartesian mind/machine dichotomy. My theory suggests that motor rhythm is, like gravity, a weak but pervasive force, and recent research has begun to also associate dysmetria with conditions like dyslexia (Nicolson et al. 1995; Ivry and Keele 1989).

Autosomal dominance

Rather remarkably, the distribution of symptoms found throughout three generations of Gopnik's subject family quite conclusively implied that the disorder was *autosomally dominant* (Hurst et al. 1990). Similarly, Kripkee and others have reported evidence of autosomal dominance for dyslexia. There is also evidence to suggest autosomal dominance in Williams syndrome, and Tourette syndrome is known to be autosomally dominant. Autosomal dominance means that a genetic trait is neither sex-linked nor recessive. Other things being equal, it means that these language disorders should become increasingly prevalent in the human gene pool!

Put differently, it seems likely that we are all language-impaired. To be sure, some of us are more impaired than others. As we noted in the case of Tourette syndrome, disorders can exhibit incomplete penetrance: pre- and postnatal factors can partially block expression of even a dominant trait. In most cases, I believe our brains simply *learn* to compensate for our disabilities. We may need to learn as much to survive our disabilities as we need to learn to survive our environment.

## Postnatal Language Learning

Despite the gauntlet of teratogens, diseases, and genetic anomalies we encounter, we mostly survive, and by birth much remarkable neural development has occurred. After the augmentation of the planum temporale, the arcuate fasciculus and auditory pathways from ear to cortex have been laid down and have begun to myelinate.[10] By about five months, the child's full complement of neurons has been produced by mitotic division. No more will be created. For the next few years, many of these cells will continue to grow and form connections among themselves, but at the same time many of them, perhaps even most of them, will begin to die off in a massively parallel competition to the death.

This phenomenon is sometimes discussed along with "programmed cell death," or *apoptosis*, the theory that every cell of the human body is programmed to die. Apoptosis is an especially important theory in areas like oncology, where it explains tumors as a programmatic failure that makes cancerous cells "immortal." In the case of neurons, however, it seems that this early cell death is less programming than the result of neurons failing to make the right connections; I will call it *neuroptosis*.[11]

Much of what we know of human neural connectivity was learned when the injury and death of a first population of neurons (as in amputation) caused atrophy and death in a second population which it innervated. Such atrophy-and-death by inactivation was the technique used in Hubel and Wiesel's early studies, and it would also explain a case of human neonatal neuroptosis first observed by Ramón y Cajal. He discovered an entire class of neurons in the most superficial layer of fetal human cortex ("horizontal cells") that are not present in adult cortex. These cells' dendrites are oriented parallel to the cortical surface (i.e., horizontally), giving the appearance that their outward developmental migration ends in a "crash" against the skull. Crashed against the skull and misoriented for resonance in cerebral cortex, where neurons are optimally arranged in vertical barrels, they atrophy and die.

Neuroptosis was first well documented in modern science in chick embryos, where up to 50% of the spinal cord neurons were found to die off between six and nine days after fertilization (Hamburger 1975). Similar neuroptosis was then rapidly found in other species, with Lund et al. (1977) finding a regression in the spininess of primate (macaque) pyramidal neurons between age eight weeks and two years. It is now clear that a large-scale sacrifice of neurons occurs as well in the human fetus and then continues at a decreasing rate for some months and years after birth. It is as if the fortunate neurons get into college, get connected, and thrive; the unconnected lead marginal lives and die young.

The child's first year is not totally prelinguistic. Considerable receptive language learning clearly occurs, but the infant's capacity for motoric response is limited, and only the most careful and ingenious experiments can assay the extent of this receptive learning. In one such experiment, Jusczyk and Hohne (1997) read 15 eight-month-old infants three stories containing novel words ten times over a two-week period. At the end of the period, the infants showed a small but statistically significant tendency to turn their heads in attention to words which they had heard in the stories.

Mothers know their children are listening, and their speech is also carefully tuned to the developing infant's language-learning needs. In chapter 6 we noted that the mother's and child's voices have poorly defined vowel formants because they have fewer harmonics than adult male voices, and we wondered how this could be adaptively conducive to language learning. It turns out, however, that "motherese" and "baby talk" are especially characterized by precise vowels (Bernstein-Ratner 1987) and wide intonation contours (Fernald and Kuhl 1987). Thus, as mother's fundamental frequency swoops up and down in careful but expressive intonation, the harmonics also "swoop" through the vocal tract filter, filling formant resonances at *all* frequencies, not just at the discrete harmonics of a monotonous voice.

In the first year of life, Piaget found motor intelligence developing from *circular reactions:* the child can see her hand moving, and from this observation, she learns that she can move her hand.[12] From this, in turn, develops a sense of self and an evolving intelligence, leading through "egocentricity" to the social awareness of the adult.

Grossberg (1986; Grossberg and Stone 1986) applied this Piagetian "schema" to babbling and the ontogenesis of speech (figure 12.3). In this model, infant babbling is just a random motor movement until some one such motor gesture (e.g., the babble *mamamamama*) becomes regularly paired with a sensory pattern (e.g., the visual presence of Mama). In figure 12.3, this occurs between $\mathcal{F}_S$ and $\mathcal{F}_M$. Adaptive grammar associates this pathway with the arcuate fasciculus. The motor pattern and the sensory pattern resonate across the arcuate fasciculus in short-term memory until, by equations 5.3 and 5.4, a long-term memory of the resonance forms.

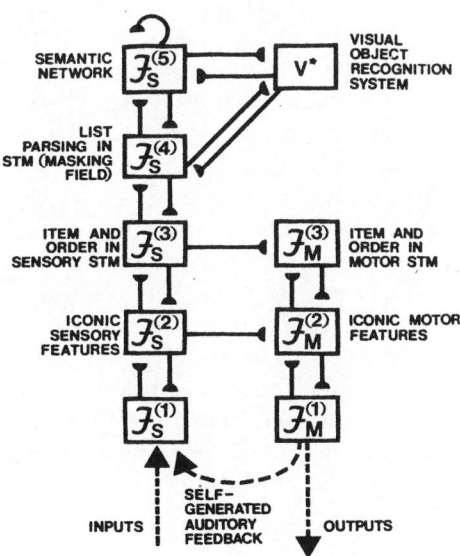

**Figure 12.3.** The Grossberg model of babbling. Babbling creates "self-generated auditory feedback," a Piagetian "circular reaction" which is mapped from sensory to motor cortex at $\mathcal{F}^2$ and learned at $\mathcal{F}^3$ and above (Grossberg 1986. Reprinted by permission of Academic Press.)

Children walk before they talk

As we noted in chapter 9, Jakobson (1968) suggested that something like *mamamamama* is universally children's first word.[13] The next universal of language learning is that children walk before they talk. Human children finally begin to walk at about one year of age. Until that time, there are a few "words" like *mamamamama* and *dadadadada*. But just after children start to walk, *mamamamama* becomes *Mama*, and *dadadadada* becomes *Dada*. Finally, at or just before age two, there is an exponential explosion of language. What triggers this sudden burst of language, if not walking?

The rhythm of walking entails a rhythmic dipole, which entrains and segments perseverative babblings into metrical feet: *mamamamamamama* becomes *Mama*, in the manner suggested by figure 9.2. Babbling gets rhythm and becomes speech. Of course, even paraplegic children can talk: phylogenetically the bilateral dipole extends back to the fishes and the bilateral brain. Even in the crib, infants exhibit vestigial tendencies to bilaterally organized motion, and crawling is also a bilateral movement.[14] But walking involves a massive dipole. When we walk, the left foot swings forward, right arm swings back; right foot swings forward, left arm swings back. A large bulk of brain becomes entrained in the rhythm of moving our bulk, even when we are babies. The neurons controlling speech articulators are not islands. They are swept up in this global undulation. And as every parent knows, when children start to walk, they start to plan. Walking rapidly becomes purposive. Toddlers start to plan how to get in trouble.

Just as the tip-of-the-tongue (TOT) phenomenon (chapter 9) revealed rhythmicity to lie near the center of word representations, so we should expect vocabulary growth to develop with walking. The child's words come slowly at first in sucking-rhythm syllables gathered into perseverative breath-group babbles like mamamamamama. Then another rhythm is added, and the syllables organize themselves in paired feet like *Mama* and *bye-bye*.

Now the child is introduced to literature in the form of rhythmic nursery rhymes, *Kinder- und Hausmaerchen*. Parents begin to teach children to say not just *bye-bye* but also *bow-wow* and *moo-cow*, and soon the child's syllables are no longer just being reduplicated; they are being combined. As Branigan 1979 showed, two word utterances like *Car go* are now being uttered in the same rhythmic time frame as the child's previous one-word utterances. As the child begins to run and her speech becomes fluent, there is an explosive growth in vocabulary. By the third year, the child may be learning new words at the rate of ten per day (Miller and Gildea 1987). This marks the beginning of morphology and syntax as on the offbeats the child begins to balance nouns and verbs with the *-ing*s and *-es*'s of fluent, grammatical speech.[15]


Imitation

Forty years ago, it was easy for Skinner to explain language learning. The child, who simply "imitated" adults, was positively reinforced for a good imitation and negatively reinforced for a poor imitation. In this manner, good language habits were established, and bad language habits were extinguished. As the generative network recruited linguists to the attack on behaviorism, it was widely reported that children do not imitate adults' speech. One famous dialogue was transcribed by Braine (1971) and has been widely quoted ever since as evidence against imitation (e.g., Fromkin and Rodman 1974; Pinker 1989, 1994).

> Child   Want other one spoon, Daddy.
> Father  You mean, you want "the other spoon."
> Child   Yes, I want other one spoon, please, Daddy.
> Father  Can you say "the other spoon"?
> Child   Other . . . one . . . spoon.
> Father  Say . . . "other."
> Child   Other.
> Father  Spoon.
> Child   Spoon.
> Father  Other . . . spoon.
> Child   Other . . . spoon. Now give me other one spoon?

And then there was the dialogue from McNeill 1966 that we quoted in chapter 1:

> Child   Nobody don't like me.
> Mother  No, say "Nobody likes me."
> [seven more times!]
> Mother  Now listen carefully. Say "Nobody likes me."
> Child   Oh, nobody don't likes me.

Such anecdotes illustrate many points about the language learning process, but the irrelevance of imitation is not one of them. As Fromkin and Rodman properly noted, parents do not often correct children as in the preceding dialogues. Absent the watchful eye of a psycholinguistics researcher's camcorder hovering over the parent-child dyad like a censorious Big Mother, parents are usually concerned more with the meaning than the form of their children's speech. They usually do not try to teach something the child is not ready to learn. To make this point more plainly, consider the following, hypothetical dialogue:

> Child   Want other one spoon, Daddy.
> Father  Where are the Himalaya mountains?
> Child   Yes, I want other one spoon, please, Daddy.
> Father  Tuesday is National Kumquat Day.
> Child   Other . . . one . . . spoon.
> Father  Pass the Chateaubriand, please.
> Child   Other.
> Father  You have a leak in your radiator.
> Child   Spoon.

> Father   I do solemnly swear to tell the whole truth.
> Child    Other . . . spoon. Now give me other one spoon?

Now this dialogue is patently absurd, and what is most absurd about it is that the father's speech at no point resonates with the child's speech. There is no topical continuity.

Fortunately, there are in the annals of child language only a few cases of such pathological parenting. In the notorious case of "Genie" (Curtiss 1977), a young girl was kept sequestered in a closet until the age of thirteen. Such "wolf children," who are not exposed to natural language until late childhood, seem unable to learn language normally in later life. The generative explanation was that, because language is innate, children do not need to be taught it, they need only to be "exposed" to language during a critical period of childhood. Wolf children, the generative model explained, simply failed to gain exposure to language during their critical period.

But as we have seen, "critical periods" seem mostly to occur before birth, and as the preceding absurd dialogue illustrates, more is needed than simple "exposure": Genie was exposed to language through the closet door, but she did not learn language. Neither do hearing children of deaf parents do not learn spoken language by watching television (Sachs et al. 1981). Because language is learned in resonance with behavioral plans, exposure alone is not enough. Language must have meaningful consequences in the social and physical environment. Consider instead a different hypothetical dialogue:

> Mother   What's this?
> Child    Koo.
> Mother   That's right! It's a cow. And what does a cow make? (pointing)
> Child    Mik.
> Mother   Yes! It makes milk.

From this much more typical dialogue, we see that both the behaviorists and their generative critics had the imitation game backward. Of course, children cannot imitate their parents. You might as well ask me, a pathetic neophyte pianist, to imitate a Horowitz recording. Rather, it is *parents* who imitate *children!* The mother's *expansions* or *recasts* of the child's utterance in the preceding dialogue are characteristic of caregiver/teacher speech and stand in stark contrast to the atopical preceding dialogue.

Child language researchers found that parents almost never gave children explicit grammatical corrections or judgments, and a branch of generative philosophy known as *learnability theory* (Gold 1965, 1967; Wexler and Culicover 1980; Berwick and Weinberg 1984) argued that language could not be learned without such "negative evidence." This formed yet another generative argument for the innateness of language. However an increasing body of research began to find expansions and recasts to be an effective "teaching method," one that is used universally by first-language caregivers as well as second-language teachers (Cross 1978; Barnes et al. 1983; Bohannon and Stanowicz 1988; Bohannon et al. 1990). These researchers have taken the general position that expansions and recasts constitute speech acts of *implicit* negative evidence or "unconscious" learning (Schmidt 1993, 1994) which vitiate the argument of learnability theory. Adaptive grammar concurs with this analysis, but it finds that the negative feedback is more deeply implicit still.

When the child says [ku] and the mother says [kau], what the mother says will resonate with the child's motor plan across the child's arcuate fasciculus, in much the same preconscious fashion as a circular reaction (figure 12.3). But *only* what *both* the mother and the child say *alike* will resonate: only what is *grammatical* will resonate. In our example, only the [k] and the [u] will resonate. What the child says incorrectly will not resonate: in our example, the child's [k-u] formant transitions will not resonate with the mother's [k-a] formant transitions.

It is not necessary for the mother to say *No, that's wrong! It's a [kau]*. The inhibitory surrounds of cerebral cytoarchitecture provide "negative evidence" for free. In parental expansions or recasts like *That's right, it's a cow*, correction need not be overt; it is automatic. If, as generative philosophy would have it, the function of a teacher could be reduced to only saying No! whenever a mistake is made, then as surely as birds learn to fly without a teacher, it would be true that children learn language without a teacher.

## No!

But *No!* is a powerful word, even if parents and teachers don't *have* to use it. In chapter 10, I suggested that *No!* unleashes nonspecific arousal which can rebound active dipoles. This nonspecific arousal may be nowhere more evident than in the screamed *No!* that heralds the advent of "the terrible twos." If the first word the child learns is mama, then the second word is *No!* Subsequently, the child begins to use *no* in combination with other words to rebound more specific conceptual networks, as in "pivot grammar"[16] sentences 12.2–12.4 (Bloom 1970):

| | |
|---|---:|
| No 'chine. | (12.2) |
| No more. | (12.3) |
| No more noise. | (12.4) |

These forms are then often succeeded by forms like 12.5–12.9.

| | |
|---|---:|
| No Fraser drink all tea. | (12.5) |
| No put in there. | (12.6) |
| Don't want baby. | (12.7) |

Allgone milk. (12.8)

Nobody don't like me. (12.9)

The multiplicity of negative forms (*no, don't, allgone, nobody,* etc.) in English obscures general patterns, but there does seem to be a tendency for an emphatic and nonspecific NEG to assume a primacy position in the child's early syntax (Bellugi 1967). Only later does the NEG become reordered, albeit often still imperfectly realized, into the fluent rhythm of standard English syntax, as in 12.10–12.12:

Fraser no drink all tea. (12.10)

Milk allgone. (12.11)

I'm not a little girl; I'm a movie star. (12.12)

In chapter 10 I followed Ross in analyzing *NEG* as being applied to the rightmost element of the sentence. I noted that, if this were true, it would imply that, contra a central tenet of adaptive grammar, there *is* a pushdown-store automaton somewhere in the brain. In the preceding examples one can see what was wrong with that analysis: what is negated is not the rightmost element of the sentence containing *NEG*. What is negated is the rightmost element of the *preceding* sentence, of the preceding *conversational turn.* Eventually, negative morphemes are encoded in the offbeat relation gradient (sentences 12.10–12), but as sentences 12.2–9 show, negation is fundamentally not a morphosyntactic phenomenon. Negation is primarily—and ontogenetically—a *discourse* phenomenon. It is the rejection of the topicalization of new information.

Syntax

Syntactic evidence has long been the foundation of the generative claim that language is not learned. We saw in chapter 10 that sentences of considerable complexity can be readily accounted for by the general mechanisms of adaptive resonance theory, without special appeal to innate homunculi. Nevertheless, it might be well to consider one final example, a classic line of argument that Pinker (1989) calls "Baker's paradox" (Baker 1979):

Irv loaded eggs into the basket. (12.13a)

Irv loaded the basket with eggs. (12.13b)

Irv poured water into the glass. (12.14a)

*Irv poured the glass with water. (12.14b)

Sentence 12.13a supposedly allows a "locative movement transformation" and admits 12.13b, but 12.14a does not admit *12.14b. The questions asked are (1) how do children come to produce sentences like *12.14b, which they do, when they never hear adults speak such sentences;(2) how do children come to *stop* using such *overgeneralized* solecisms if adults never correct them; and (3) since adults never correct them, how do children distinguish such valid and invalid constructions, whose verbs seem otherwise synonymous? Baker's paradox leads to some amusing syntactic puzzles, but like the puzzles created by the generative deduction, they are based on unwarranted premises; namely, (1) that something "moves" and (2) that this "movement" is governed by logical, computer-like rules.

In general, instead of insisting that language is a rule-governed, computer-program-like system, I assume that language is learned by brain cells. Then children can overgeneralize and produce patterns like *12.14b because cerebral competition has not yet contrast-enhanced and partitioned their linguistic concepts. Just as a child can call a cow a *doggie*, a child can say *I poured the glass with water*. Eventually, children learn that there are many different verbs which admit many different case frames. Children can make these many and fine distinctions because the massively parallel architecture of their cerebrum "computes" these patterns with a granularity approaching 1 part in $10^{7}$[111,111]. Discrete rules like "locative movement" fail to account for 12.15–12.17 because they deny that language can be complex to this degree. Sentence *12.15 illustrates this complexity with the verb *to fill*, which does not admit a locative case role in the first place.

*Irv filled water into the glass. (12.15)

Irv poured the glass with ice water. (12.16)

The waiter poured the glasses with Chateau Petrus 1961. (12.17)

Sentences 12.16 and 12.17 raise paradoxes within Baker's paradox. If we assume language is rule-governed behavior, not only must we explain how children *stop* saying sentences like *12.14b, but now we must also explain why adults, once having "acquired" the rule, then go on to violate it with sentences like 12.16 and 12.17. The problem with rule-governed explanations of language has always been that the rules are more observed in the breach than in the observance.

The brain is complex, more complex by far than I have made it seem in these pages. Even so, it seems simpler to allow that language is adaptive: we grow, we change, we learn, and our language grows and changes with us. So, for example, children stop overgeneralizing to forms like *12.14b when they learn that *water* becomes a kind of default patient of the verb *pour:* water is what one *usually* pours. In *12.14b, *water* is hardly the sort of thing one would explicitly present as new information. In 12.16, however, *ice water* could be new information. More clearly still, in 12.17 or in any context, Chateau Petrus 1961 would definitely be rare and new information. As we adapt to new information, our grammar must adapt with us.

Reading

By the time the child goes to school, a great deal of language has been learned. The basic motor plans of language have been laid down in cerebral cortex and their rhythms coordinated through cerebellar learning. Many verb patterns have also been learned and associated with many appropriate nouns and case roles. A basic inflectional grammar has also been learned, a neural network which inserts *copula* and *-en* into English passive sentences and in a thousand other particulars maintains an offbeat grammatical commentary on the semantic substance of the sentence. For the most part, the syntactic order of nouns, verbs, and their modifiers need not be learned. These parts of speech follow a universal topic primacy gradient which is innate, but which is neither language-specific nor species-specific.

In school, reading is the pupil's first task, and in English it is notoriously difficult. Until Chomsky and Halle's *Sound Pattern of English* (1968), a considerable amount of research was devoted to the sound-spelling correspondences of English, many of which are quite irregular. This research often led to some rather fanciful theories. The most infamously entertaining illustration of these was probably G. B. Shaw's spelling of *fish* as *ghoti*, using the *gh* of *enough*, the *o* of *women*, and the *ti* of *nation*. Linguists were quick to point out that, although the *o* of *women* is idiosyncratic, Shaw had ignored the fact that the other phoneme-grapheme correspondences were context-sensitive: *gh* only assumes the sound of *f* in syllable-final position and *ti* only assumes the sound of /ʃ/ before a following vowel. These observations point out the fact that reading is context-sensitive, but they largely ignored the fact that it is also rhythm-sensitive, as in the *reciprocate/reciprocity* contrast cited in chapter 8.

Reading is a double serial process. In the first place, the serial array of the written word must be visually processed. As we saw above, one type of dyslexia may affect the cerebellar control of eye saccades, which subserves this process. Then, the visually scanned information must be associated with one or several words—phonological motor maps.[17] A second type of dyslexia could impede this association, and this type of dyslexia may in fact be *induced* by instruction.

In English, there is a significant tendency for many poor and "dyslexic" readers to "plateau" around the age of ten. One clue to a possible cause of reading problems at this level in English comes from Holmes and Singer (1966), who found that at this age, knowledge of Greek and Latin roots was a significant predictor of subsequent reading achievement. It is mostly only English polysyllabic words which have Greek or Latin roots, and it is mostly only these same words which exhibit stress alternations in English. Given the implication of stress patterns in lexical retrieval by the TOT phenomenon, one suspects stress alternations might be implicated in this reading problem. Indeed, many poor and/or dyslexic readers can be heard to attempt to read a word like *official* as "off . . . awfick . . . often . . . awful."

Part of the student's problem here is that he is trying to do exactly what his teachers have told him to do: he is trying to sound out the word from left to right. But the English stress patterns are predicted by the *end* of the word (Chomsky and Halle 1968; Dickerson 1975; Dickerson and Finney 1978). In words like *official,* failure to first find the right stress pattern can entail failure to find the right initial segment. Without the stress or the initial segment, the TOT phenomenon predicts the student will be unable to find the word he is looking for. Instead, the student should be decoding polysyllabic words from *right to left!* Other factors in reading (e.g., socioeconomic status of the student, the intrinsic interest of the reading material) may be more significant to successfully learning to read, but the rhythmic integration of syllables into feet and words should be no less important to reading than it is to speech, and the failure of reading theory to integrate syllables and whole words in this fashion may be one reason for the inconclusiveness of the perennial debate between "phonics" and "whole-word" approaches to English reading instruction.

Once the pupil can use printed words to access words in his mind—by sound, by meaning, *and* by rhythm—whole other worlds of vocabulary open up. Indeed, whole other languages open up. There is the language of geography, the language of history, the language of biology—all the literatures of letters and jargons of science, all the languages of the world that use the Roman alphabet. (The languages written in other scripts, Korean in *hangul,* Hindi in *devanagari*—not to mention mathematics—are quite another matter, as are the nonalphabetic scripts of Chinese and Japanese.) Reading is fundamental, but it is not parochial, and it is a learning activity which extends well into adulthood.

## Adult Learners

Pronunciation is the aspect of language in which adult second-language learners most frequently fail to achieve the proficiency of young language learners. In learning the pronunciation of a second language, adults encounter five difficulties which must be explained.

The first is in some ways the easiest. Adults often simply don't hear the difference between two sounds. This disorder appears to begin in adolescence, when children begin not to hear when their parents ask them to do something, and it continues into adulthood. In chapter 7, I discussed a Spanish bilingual who distinguished /ʃit/ and /ʃit/ by tone of voice instead of by vowel formants. Adults have already learned a thing or two, and they can figure out what words mean without figuring out exactly how they sound.

The second difficulty is in pronouncing sounds which are *different* from those of the adult learner's native language. For example, German does not have an equivalent of the English [w], and when *w* is written in German, it is pronounced as [v]. Most German speakers have difficulty pronouncing *w* as [w]; they pronounce it as [v] in English as they do in German. The third type of difficulty, first-language *interference,* occurs when pronouncing L2 sounds which are *similar* to L1 sounds (Flege 1988). As an example of this difficulty, we might cite the /ɪ/-/i/ contrast discussed in chapter 7. As noted in that chapter, these difficulties were paradoxical under behaviorist assumptions but can be substantially reconciled in terms of cerebral phonological motor maps constructed over on-center off-surround anatomies: in such anatomies similar sounds can *mask* each other.

The fourth problem is difficulty pronouncing the same problem phones L1 children have difficulty pronouncing. Examples of these problem phones include English /r/ and /I/. The fifth problem is maintaining accurate pronunciation in rapid speech. L1 problem phones tend to require "nonquantal" articulations which, like rapid speech, require exceptionally fine motor control. Adaptive grammar suggests that rapid and problem phones are difficult for adult learners because the phones require fine cerebellar learning. As we saw in chapters 4 and 5, the cerebellum lacks the on-center off-surround architecture of the cerebrum and is therefore incapable of rapid, rebound-mediated learning. Consequently, both the L1 child and the L2 adult learn these pronunciations slowly.

Adults' capacity to master these last two categories of pronunciation also appears to decrease with age more rapidly than their capacity to master the first two categories. Insofar as this is true, it appears to be akin to the adult's decreasing capacity to learn fine motor skills like playing the violin, so it seems misleading to attribute the child's advantage to some innate, language-specific capacity. Moreover, the decrease in these capacities appears to be relatively linear, so it also seems misleading to characterize it as the result of some "critical period." Adaptive grammar attributes the child prodigy's advantage to the fact that the cerebellum is a tabula rasa at birth and a slow, nonplastic learner thereafter.

Because basic syntax does not directly involve the cerebellum, it is normally learned much better than pronunciation, but insofar as L2 inflectional morphology involves sound and rhythm, it depends more heavily upon cerebellar processes and presents an intermediate degree of difficulty. Failure to completely control morphology is often called *fossilization* (Selinker 1972), an apt term within adaptive grammar, implying as it does the long-term memory calcification of postsynaptic membranes.

## Writing

After reading, there is writing. At first, this is simply "penmanship," a fine motor skill requiring endless drill in order to train the punctilious cerebellum. But writing—and the more inclusive term, literacy—have consequences far beyond penmanship. Tannen's book *Spoken and Written Language* (1982) is a seminal analysis of the systematic stylistic differences between spoken and written communication; other authors (e.g., Ong 1982; McLuhan 1965; Olson 1991) have also called attention to the broad, cultural consequences of literacy. Here I would note only that in the family and in the neighborhood school, the pupil's language, whether spoken or written, tends at first to be parochial. The words she uses are the words of the caregiver and teacher. But once a word is committed to writing, it lives beyond the moment, beyond the neighborhood, beyond the time when its network can be modified, modulated, or even rebounded by immediate feedback from an interlocutor. The written word acts on a universal and eternal stage. Once the pupil learns this fact, writing becomes more than a motor act, it becomes a public behavior in the world and a public behavior in history.

In the limit, the activated network which wills an act of writing must be planned in the context of all the competing networks of all possible readers that the author can imagine, in the context of all consequences of conceivably practical bearing. If it is in reading that students first really sample the diversity of other minds, other times, and other cultures, then it is in writing that this learning becomes behavior. In writing, the child achieves a kind of sociosynaptic maturity and becomes a publicly communicating node in the society of *Homo loquens*. In the task of writing to a pluralistic and changing social environment, the student learns to say *No!* to the parochial, to rebound her expectancies, and to imagine other cultures and contexts where meaning might be found.

# NOTES

CHAPTER 1

1. Aristotle, *De partibus animalium* 2.7.
2. See DeFelipe and Jones 1988 for a recent appreciation of Ramón y Cajal's work.
3. The force of this argument still holds over a wide range of assumptions. Estimates of the number of human neurons vary between $10^{10}$ and $10^{12}$, so Jackendoff's $10^{10}$ is reasonably conservative. While 1.8 overestimates ν by implying that two subnetworks of neurons differing by only one connection can encode distinct representations (a most unlikely case), it also underestimates ν by disregarding orderings among the synapses. If order were important, and were computed as permutations, it would be larger by a factor of $k!$ Order is important, but it is not that large a factor. In general, $k$ can vary by ±4 orders of magnitude and still $\nu \gg 10^{1111}$. Throughout the text, I estimate ν by $10^{7,111,111}$ to emphasize the fact that ν is big.
4. As just one example of this presumption, consider Simon and Feigenbaum 1979: "[we assume that] the central processing mechanism operates serially and is capable of doing only one thing at a time."
5. The story bears repeating. Einstein could not abide Heisenberg's conclusion that wave-particles could not be definitively measured but only statistically estimated. "Nonsense, Niels!" Einstein exclaimed. "God doesn't play at dice!" "Albert!" Bohr replied. "Stop telling God what to do!"

CHAPTER 2

1. On the evidence of electroencephalograph (EEG) data, Poizner et al. (1987) concluded that ASL is organized in the same cortex as spoken language, but recent functional magnetic resonance imaging (fMRI) evidence points to a much greater involvement of visual and right-hemisphere cortex in sign language than in spoken language (Neville and Bavelier 1996, see also Kimura 1988).
2. I found this account of Miller's experiment in Isaac Asimov's *Beginnings* (1987).
3. As has been often remarked, the cockroach has been around a good while longer than *Homo loquens* and shows every sign of being around a good while longer.

One should not mean disrespect when using terms like "lower phyla"; "higher phyla" may not get the last laugh.

4. These are analogous to proprioceptive stretch receptors in the vertebrate locomotor system.

5. The first vertebrates in the fossil record, the fishy ostracoderms, appeared in the late Cambrian period.

CHAPTER 3

1. A few classes of neurons have short axons, but these are very much the exception.

2. In a sense, every neuron is an interneuron, since every neuron communicates between two points. Nevertheless, the term is loosely used to locally distinguish neurons that are not major input or output neurons.

3. As of this writing, nearly 1,000 human G-protein receptors have been reported in World Wide Web databases. Fortunately for neuroscience, which is already complicated enough, most of these are not CNS receptors; unfortunately, CNS receptors still number in the hundreds.

4. The principal non-NMDA receptors are quisqualate and kainate receptors.

5. There are also *autoreceptors* for serotonin and other messengers. These respond to the cell's own released neurotransmitters.

CHAPTER 4

1. In recent years the term "neocortex," or simply "cortex" (from the Latin for "rind") has come to be preferred to "cerebrum." Unfortunately, the cerebellum also has a "rind," but its cortex is of an entirely different neural design from that of cerebral cortex, and for a clear understanding of language, it is necessary to understand the differences between the two architectures. To keep these differences in the forefront, I will persist in using the older term, "cerebrum."

2. This pyramidal tract was originally named, not for the pyramid-shaped pyramidal cells which were later found to give rise to it, but rather for the pyramid-shaped gyrus of the medulla from which the spinal nerves descend.

3. Given three squares and an ordered stack of playing cards on Square 1 (ace on top, 2, 3, . . . J, Q, K), the Tower of Hanoi task is to re-create the ordered stack on Square 3 by moving one card at a time to another square, never placing a higher card on a lower card. The solution is recursive, and since recursion is a powerful computational tool, a generation of researchers took this otherwise trivial puzzle to be cognitively significant.

Damasio et al.'s 1996 finding that proper names localize to the left temporal pole, proximate to the hippocampus, also suggests that HM's particularly dramatic inability to remember even his doctor's name may have been due not only to resection of the hippocampus but also to resection of adjacent brain regions.

4. As we saw in chapter 3, matters are more complicated still. Different postsynaptic receptors can respond differently to a single neurotransmitter. Thus, for example, the ($\alpha_1$-adrenergic noradrenaline receptor is excitatory, admitting $Ca^{2+}$ into the cell interior, but the ß-receptor is inhibitory, suppressing the cAMP second-messenger system (Shepherd 1997).

5. Most aphasias present a complex of deficits, and "pure" aphasias of one sort or the other are rare.

6. "Stellate" has been used to describe any vaguely star-shaped cell, whether inhibitory or excitatory; here I use it to describe only excitatory cells. Chandelier cells did not stain well in early preparations and were not positively identified until Szentágothai himself did so in 1974.

7. A "bell-shaped curve" is an approximate example of a Gaussian probability distribution. The "waves and troughs" of these synaptic distributions are exemplified in figure 5.4b.

CHAPTER 5

1. This and related chromatic "optical illusions" are known in the literature as the *McCollough effect* (McCollough 1965).

2. Note that besides cessation of input and NSA, a rebound could also theoretically be effected by momentarily increasing or decreasing the inhibitory effects of $i_{gr}$ and $i_{rg}$.

3. Technically, chloride (Cl⁻) channels can still be open and further hyperpolarize a membrane after all Na⁺ channels are closed, but this process is limited, too.

4. As we shall see, there are two especially important exceptions to the principle of reciprocal, resonant connectivity: inhibitory cells cannot resonantly excite each other, and the cerebellum (being principally inhibitory in its Purkinje cell output) is not resonant.

5. It should be noted that the ART computer simulation presented in this chapter is derived from Grossberg's earlier, more psychological work. Subsequently Carpenter and Grossberg developed and popularized a number of computer algorithms called ART-I, ART-II, etc. (Carpenter and Grossberg 1987; Carpenter et al. 1991). To achieve efficiency, these algorithms abstract away the biological details of Grossberg's original ART. Our interest here is more in biological fidelity than computational efficiency, so the reader should not expect to find a detailed correspondence between our models and the later ART models.

6. In vivo, ocular dominance columns become convoluted through interactions with orientation maps of the selective response of striate cortex barrels to lines in the vertical-horizontal plane. See Grossberg and Olson 1994 for a detailed ART model of this interaction.

7. Technically, Minsky and Papert only claimed to have proved that one class of simple perceptrons was incapable of computing XOR. However, they used this result to argue more broadly against all parallel computational models.

CHAPTER 6

1. "Vocal fold" is a more descriptive term for these liplike organs, but idealizing them as "vocal cords" helps illustrate the acoustic principles behind voiced speech.

2. Cycles per second, abbreviated cps is now usually expressed in hertz (Hz) in honor of Heinrich Hertz, the German physicist who pioneered classical wave theory.

3. Texts in acoustic phonetics normally denote formants using F plus superscript, but the neural-modeling literature uses $F$ with superscripts, like $F^1$, to identify neuron fields. We therefore use F plus subscript to denote formants (e.g., $F_1$).

4. Vowels are also described with the tongue-place features of high, middle, and low and of front, central, and back.

5. In English and many other languages, /g/ and /k/ have a variant articulation, a palatal pronunciation, in which the tongue is positioned further forward. This

*allophone* typically occurs when /g/ or /k/ precedes a front vowel like /i/. Its formant transitions have a different spectrographic signature from velar /g/ and /k/.

### CHAPTER 7

1. The effects discussed in this section have also been discussed as a "perceptual magnet effect" (Kuhl 1991) or an "anchor effect" (Sawusch and Nusbaum 1979; Sawusch et al. 1980). See also Flege 1988 and Rochet 1991.

2. A number of these issues surrounding NSA and bilingualism are addressed in Rose 1993.

3. As we will see in chapter 12's discussion of Kim et al. 1997, bilinguals who learn their second language later in life can apparently also recruit wholly distinct $A^3$–level polypoles for the second language.

### CHAPTER 8

1. In this chapter we use $z_{ji}$ instead of $z_{ij}$ because we are mainly concerned with feedback, top-down signals from some higher node $xj$ to some lower node $x_i$.

2. If a learner is prepped to "memorize the following (long) list," he will normally focus on the beginning of the list, and a primacy effect can result, but very long lists, or lists for which the learner is not primed, will exhibit pronounced recency effects.

3. Recall from chapter 5 that the 0.5 mm inhibitory radius of an individual inhibitory cell also agrees well with the average width of ocular dominance columns in visual cortex reported by Hubel and Wiesel (1977).

4. See Bullock et al. 1994 and Fiala et al. 1996 for detailed ART models of cerebellar learning.

5. *Ataxia* is the inability to coordinate voluntary movement. *Dysarthria* is the inability to articulate words, as distinct from the ability to plan or comprehend words, which is termed *aphasia*.

6. The Kent et al. (1979) model differs from my account in positing short, inhibitory, cerebellar loops which, if handicapped by disease, force the system "to rely on longer loops to control movement. Consequently segment durations in speech are increased to allow time for the longer loops to operate."

### CHAPTER 9

1. Dolphins and whales, while not exactly bipedal, do also move with a two-beat rhythm of their flukes. Similarly, parrots and other birds move with graceful two-beat wing strokes, if not with graceful two-beat footsteps.

2. Some spectrographic studies (*e.g.,* Delattre 1965) have disputed this claim for French, and there is some evidence of stress-timed meter in French poetry (e.g., the classical alexandrine). Chinese has been widely taken to be an "isolating" language of essentially monosyllabic structure. However, a close study of the Chinese lexicon shows that most Chinese words are in fact multisyllabic. In particular, Chinese *cheng-yu* (loosely, "proverbs") assume a canonical, four-beat structure divided into two feet of two beats each. The analogy of this structure to "beats" (as in music) or "stress" (as in English) is strengthened when one considers also the Chinese tone alternations which occur on the offbeats. Where English words have "stress patterns," Chinese words have "tone patterns," and just as English offbeats undergo vowel reduction, as in *reciprocal-reciprocity*, Chinese offbeats undergo reduction to "neutral tone" (Wu 1992).

3. There are instances of stressed inflections and grammatical morphemes, for example, in Russian and Turkish. In many cases, these are only stressed where the inflection carries propositional content as well as grammatical content.

4. Generative grammar called such linguistic regularities as 9.12 "rules." Instead of *rule*, which implies serial processing, as by a Turing machine, adaptive grammar prefers to say that regularities like 9.12 actually reflect *patterns* of neural resonance.

5. However, there are occasions on which children will overgeneralize, producing plural forms like *mens* or *childrens*.

CHAPTER 10

1. We will see in chapter 11 that sentences 10.1 and 10.2 normally do *not* mean the same thing.

2. In LISP, *dynamic scoping* means that a variable is "bound" and has a specific value only below a specific tree node. Dynamic scoping is roughly comparable to notions like *subjacency* or *c-command* in generative systems.

3. John McCarthy is generally credited with designing the LISP computer language, but McCarthy himself credits Chomsky with convincing him of the value of a rigorous implementation of Church's lambda calculus on a pushdown-store automaton.

4. A pidgin language is a simple language that is typically used for trade and commerce and that is invented when speakers of two mutually unintelligible languages come into contact. A pidgin language evolves into a creole when mothers begin to teach pidgin to their children as a first language. In this process, creoles begin to evolve the offbeat morphology that is the eventual hallmark of "fully developed" languages.

5. See Comrie 1989 for a discussion of accusative and unaccusative languages.

6. Linguists refer to unusual or contrasting-in-context linguistic constructions as being *marked*.

7. Less centrally, a proposition may also have attached to it a number of "outer" case arguments expressing such adverbial or prepositional relations as location, time, and manner. These are often expressed or implied "outside" the sentence.

8. There are other constraints on particle movement and dative movement. For example, we can say *He threw back the ball*, but not *\*He threw back the ball to Ted*. See Fraser 1976 for a detailed analysis.

9. Generative linguistics, which recognized the existence of null pronouns in "pro-drop" languages like Spanish, was reluctant to admit them to the analysis of English. In the last several years, generative linguistics appears to have moved toward the position developed here.

CHAPTER 11

1. Some scholars have found it ironic that the epigraph to this chapter appeared in Peirce's essay "How to Make Our Ideas Clear" (1878). Perhaps the epigraph has suffered in translation from Peirce's original French.

2. Copernicus would no doubt have gotten into trouble, too, if he hadn't had the good sense to die on the day he published *De revolutionibus*.

3. There were, to be sure, earlier translations of the Bible into the vernacular. Wycliffe's translation of the Bible into English (1380–93) was hardly approved by the Church, but because of the Great Schism (which began in 1378 and at one point resulted in three contending popes), the Church did not have its act together, and Wycliffe died a natural death. However, Wycliffe's follower, Jan Huss (1369?–1415) was

burned at the stake for the very idea of a vernacular Bible, and in the same year the Church exhumed Wycliffe's body and burned it, just for good measure. Even Luther's contemporary William Tyndale was executed for his English translation (1534), which became the basis of the English King James Version. Luther's version contains some beautiful German, but he's mostly famous for getting away with it.

4. In countries where English is foreign, the term "English as a foreign language," or EFL, is used, but where English is not foreign, one obviously should not call it English as a foreign language. In contexts where English is the norm, one speaks of English as a second language, or ESL.

5. The teaching of ESL remains something of an exception. As in the days of behaviorism, the field continues to be an exceptional laboratory for the study of the human mind. Indeed, it has raised many of the critical issues addressed by adaptive grammar in this book. But the United States no longer needs to import cheap labor. As a result, English learners in the United States are stereotyped as "illegal aliens." Illegal or not, they are increasingly unwelcome and disenfranchised. Thus, in this political climate, the entire enterprise of ESL has become stigmatized. This is less true in non-English-speaking countries of the world. There, EFL has obvious practical bearing and is regarded as meaningful.

CHAPTER 12

1. Kim et al. (1997) did not find separate L1 and L2 regions in Wernicke's area. This could be an artifact of their design, but it might also be evidence in support of the single-polypole, contextual dipole model of figure 7.8.

2. Derived from the Greek *teratos* for "monster," teratology is the clinical term for the study of birth defects. Testosterone is not clinically recognized as a teratogen.

3. The association between influenza and schizophrenia, though weak and poorly understood, has been corroborated, and a further association has been found between second-trimester influenza and adult unipolar depressive disorder (Machon et al. 1997).

4. As my discussion should make clear and as Pinker (1994) explains at some length, genes do not have simple, unitary effects. Rather, they exert effects in combination with other genes, so there are probably hundreds of genes which could lay fair claim to being "grammar genes."

5. Many researchers have hypothesized that perceptual saliency is a factor in learning inflectional morphology (e.g., Brown 1973; Clahsen 1989).

6. Bound, syllabic inflectional morphemes like *-ing* present an intermediate case between syllabic forms like *of* and nonsyllabic forms like *-s*. Common instances of such inflections (e.g., *go/going*) may be learned and performed "on the downbeat" as if they were uninflected irregular forms (cf. *throw/threw* or *go/gonna*). Even dysphasics may not exhibit difficulties with these forms in normal conversation. On wug tests, however, the nonsense forms cannot be previously learned, and here the dysphasics' deficit becomes especially apparent.

7. Dyslexia is not well understood, and the classification probably includes several subtypes of different etiology. The historically dominant view, presented by Vellutino (1987), was that dyslexia is a higher cognitive disorder. Dyslexia is four to ten times more common in boys than girls, and there is reason to associate this with Geschwind's finding of lesser cerebral asymmetry in boys. Such a subtype would not be inconsistent with adaptive grammar. However, as techniques for the measurement of fine eye movement have improved, there have been increasing reports of subtle

fine motor disorders among dyslexics, although overshoots and "cerebellar braking problems" as described in the text are not always found. Raymond et al. (1988) found a positive correlation between cerebellar subtle dysfunction, dyslexia, and gaze instability, and Biscaldi et al. (1994) even reported a subtype characterized by saccadic *undershoot.* See Aral et al. 1994 for a recent model of eye saccades involving the cerebellum.

8. Relatively few chemicals pass from the blood into the brain or from the brain into the blood. This fact is called the *blood-brain barrier*.

9. The collocation of "professors and brainwashers" is intended to alert the reader to the fact that this is very much a two-edged sword, as discussed further below.

10. Myelination is another often-postulated neurological cause of a critical period for language (Long 1990). Adaptive grammar does not preclude a causal relationship between myelination (e.g, of the arcuate fasciculus) and language learning, but like lateralization, myelination is not species-specific, let alone language-specific, and it seems mostly to occur *before* the child begins to learn language.

11. See Deshmukh and Johnson 1997 for a recent study elaborating this distinction between neural cell death and apoptosis.

12. One should not conclude that a paralyzed child who cannot move her hand cannot develop cognition (although in such cases one should not be surprised to find aspects of cognition developing more slowly, either). The point is that proprioceptive feedback, of various origins, can form a "circular reaction," which can be learned and remembered at long-term memory traces for later, volitional playback.

13. Long vocalizations like *mamamamamamama* probably do not have a specific meaning associated with them. They may be more words in the mind of the mommy than in the intentions of the baby. And, of course, any one child's first word could also be *bye-bye* or *No!*—or the name of the family dog. But the early, if not absolutely first, occurrence of /mama/ is still far, far too widespread a phenomenon to be coincidental.

14. I thank Elaine Shea for pointing this out to me.

15. In English, inflectional *-s* and *-ed* are usually not syllabic, so a technical account of the learning of inflectional and derivational morphology as an offbeat process must appeal to other child language morphophonemic processes like reduplication and diminutive affixation as well.

16. Braine (1971) introduced the notion of "pivot grammar" to describe the child's early two-word utterances. Subsequent researchers were unable to generalize the notion to longer utterances, but it anticipated adaptive grammar's implication of rhythmic dipoles in the syntactic organization of language.

17. Even in Chinese, whose writing system is often thought to bear little or no overt relationship to the spoken language, we find that Chinese readers process Chinese characters primarily as sound (Chu-Chang and Loritz 1977).

# REFERENCES

Ackley, D., Hinton, G., and Sejnowski, T. 1985. A learning algorithm for Boltzmann machines. *Cognitive Science 9*: 147–69. Reprinted in Anderson and Rosenfeld 1988.
Aghajanian, G. K., and Marek, G. J. 1997. Serotonin induces excitatory postsynaptic potentials in apical dendrites of neocortical pyramidal cells. *Neuropharmacology 36* (4–5): 589–99.
Allen, G. I., and Tsukahara, N. 1974. Cerebrocerebellar communication systems. *Physiological Review 54*: 957–1006.
ALPAC. 1966. *Language and machines: computers in translation and linguistics.* A report by the Automatic Language Processing Advisory Committee of the Division of Behavioral Science, National Academy of Science, National Research Council. Publication no. 1416. Washington: National Academy of Science/National Research Council.
Anderson, J. A., and Rosenfeld, E., eds. 1988. *Neurocomputing: foundations of research.* Cambridge: MIT Press.
Anson, J. 1977. *The Amityville Horror.* New York: Prentice-Hall.
Aral, K., Keller, E. L., and Edelman, J. A. 1994. Two-dimensional neural network model of the primate saccadic system. *Neural Networks 7* (6/7): 1115–35.
Aristotle. 1908. *De partibus animalium.* Trans. J. A. Smith and W. D. Ross. Oxford: Clarendon Press.
Asimov, I. 1987. *Beginnings: the story of origins—of mankind, life, the earth, the universe.* New York: Walker.
Austin, J. L. 1962. *How to do things with words.* New York: Oxford University Press.
Baker, C. L. (1979). Syntactic theory and the projection problem. *Linguistic Inquiry 10*: 533–81.
Barnes, S., Gutfreund, M., Satterly, D., and Wells, G. 1983. Characteristics of adult speech which predict children's language development. *Journal of Child Language 10*: 65–84
Barnsley, M. F. 1988. *Fractals everywhere.* New York: Academic Press.
Bates, E., and MacWhinney, B. 1982. Functionalist approaches to grammar. In E. Wanner and L. R. Gleitman, eds., *Language acquisition: the state of the art.* Cambridge: Cambridge University Press.

Békésy, G. von. 1960. *Experiments in hearing.* New York: McGraw-Hill.
Bellugi, U. 1967. "The acquisition of negation." Ph.D. diss., Harvard University.
Berko, J. 1958. The child's learning of English morphology. *Word 14*: 150–77.
Bernstein-Ratner, N. 1987. The phonology of parent-child speech. In K. E. Nelson and A. van Kleeck, eds., *Children's language.* Hillsdale, NJ: Lawrence Erlbaum.
Berwick, R. C., and Weinberg, A. S. 1984. *The grammatical basis of linguistic performance.* Cambridge: MIT Press.
Bickerton, D. 1981. *Roots of language.* Ann Arbor: Karoma Press.
Biscaldi, M., Fischer, B., and Aiple, F. 1994. Saccadic eye movements of dyslexic and normal reading children. *Perception 23* (1): 45–64.
Blanken, G., Dittmann, J., Grimm, H., Marshall, J. C., and Wallesch, C. W., eds. 1993. *Linguistic disorders and pathologies.* Berlin: de Gruyter.
Bliss, T. V. P., and Lømo, T. 1973. Long-lasting potentiation of synaptic transmission in the dentate area of the anaesthetized rabbit following stimulation of the perforant path. *Journal of Physiology (London) 232*: 331–56.
Bloom, L. 1970. *Language development: form and function in emerging grammars.* Cambridge: MIT Press.
Bobrow, D. G., and Fraser, B. 1969. An augmented state transition network analysis procedure. *International Joint Conferences on Artificial Intelligence 1*: 557–67.
Bohannon, J. N., MacWhinney, B., and Snow, C. E. 1990. Negative evidence revisited: beyond learnability or who has to prove what to whom? *Developmental Psychology 26*: 221–26.
Bohannon, J. N., and Stanowicz, L. 1988. Adult responses to children's language errors: the issue of negative evidence. *Developmental Psychology 24*: 684–89.
Bolinger, D. 1968. *Aspects of language.* New York: Harcourt, Brace, and World.
Braine, M. D. S. 1971. On two types of models of the internalization of grammars. In D. I. Slobin, ed., *The ontogenesis of grammar: a theoretical symposium.* New York: Academic Press.
Branigan, G. 1979. "Sequences of words as structured units." Ph.D. diss., Boston University.
Brodmann, K. 1909. *Vergleichende Lokalisationlehre der Grosshirnrinde in ihren Prinzipien dargestellt auf Grund des Zellenbaues.* Leipzig: J. A. Barth.
Brown, R. 1973. *A first language.* Cambridge: Harvard University Press.
Brown, R., and MacNeill, D. 1966. The "tip-of-the-tongue" phenomenon. *Journal of Verbal Learning and Verbal Behavior 5*: 325–37.
Bullock, D., Fiala, J. C., and Grossberg, S. 1994. A neural model of timed response learning in the cerebellum. *Neural Networks 7* (6/7): 1101–14.
Cairns-Smith, A. G. 1985. *Seven clues to the origin of life.* Cambridge: Cambridge University Press.
Campos, H., and Kemchinsky, P. 1997. *Evolution and revolution in linguistic theory: studies in honor of Carlos P. Otero.* Washington: Georgetown University Press.
Carpenter, G. A., and Grossberg, S. 1987. A massively parallel architecture for a self-organizing neural pattern recognition machine. *Computer Vision, Graphics, and Image Processing 37*: 54–115.
Carpenter, G. A., Grossberg, S., and Rosen, D. B. 1991. Fuzzy ART: fast stable learning and categorization of analog patterns by an adaptive resonance system. *Neural Networks 4*: 759–71.
Chafe, W. 1970. *Meaning and the structure of language.* Chicago: University of Chicago Press.
Changeux, J.-P. 1985. *Neuronal man: the biology of mind.* New York: Pantheon Books.

Chao, Y.-R. 1948. *Mandarin primer: an intensive course in spoken Chinese.* Cambridge: Harvard University Press.
———. 1968. *A grammar of spoken Chinese.* Berkeley: University of California Press.
Chomsky, C. 1969. *The acquisition of syntax in children from five to ten.* Cambridge: MIT Press.
Chomsky, N. 1959. Review of B. F. Skinner's *Verbal behavior. Language 35* (1): 26–57.
———. 1965. *Aspects of the theory of syntax.* Cambridge: MIT Press.
———. 1971. Review of B. F. Skinner's *Beyond freedom and dignity,* "The case against Skinner." *New York Review of Books 18* (11): 18–24 (Dec. 29). Reprinted in F. W. Matson, ed., *Without/within behaviorism and humanism.* Belmont, CA: Wadsworth.
———. 1972. *Language and mind.* New York: Harcourt Brace Jovanovich.
———. 1988. *Language and politics.* Montreal: Black Rose Books.
———. 1993. *Language and thought.* Wakefield, RI: Moyer Bell.
———. 1994. Bare phrase structure. MIT Working Papers in Linguistics. Cambridge. Reprinted in Campos and Kemchinsky 1997.
———. 1995. *The minimalist program.* Cambridge: MIT Press.
Chomsky, N., and Halle, M. 1968. *The sound pattern of English.* New York: Harcourt Brace Jovanovich.
Chu-Chang, M., and Loritz, D. 1977. Phonological encoding of Chinese ideographs in short-term memory. *Language Learning 31*: 78–103.
Church, A. 1941. *The calculi of lambda-conversion.* London: Oxford University Press.
Ciuffreda, K. J., Kenyon, R. V., and Stark, L. 1985. Eye movements during reading: further case reports. *American Journal of Optometry and Physiological Optics 62* (12): 844–52.
Clahsen, H. 1989. The grammatical characterization of developmental dysphasia. *Linguistics 27*: 897–920.
Cohen, M. A., and Grossberg, S. 1986. Neural dynamics of speech and language coding: developmental programs, perceptual grouping, and competition for short term memory. *Human Neurobiology 5*: 1–22.
Comrie, B. 1989. *Language universals and linguistic typology.* 2d ed. Oxford: Basil Blackwell.
Cook, W. A. 1989. *Case grammar theory.* Washington: Georgetown University Press.
Copernicus, N. 1543. *De revolutionibus orbium coelestium.*
Crick, F., and Asanuma, C. 1986. Certain aspects of the anatomy and physiology of the cerebral cortex. In McClelland and Rumelhart 1986.
Cross, T. 1978. Mother's speech and its association with rate of linguistic development in young children. In N. Waterson and C. Snow, eds., *The development of communication.* New York: Wiley.
Crowder, R. G. 1970. The role of one's own voice in immediate memory. *Cognitive Psychology 1*: 157–78.
Curtiss, S. 1977. *Genie: a psycholinguistic study of a modern-day "wild child."* New York: Academic Press.
Cutler, A., ed. 1982. *Slips of the tongue and language production.* Chichester: Wiley.
Damasio, H., Grabowski, T. J., Tranel, D., Hichwa, R. D., and Damasio, A. R. 1996. A neural basis for lexical retrieval. *Nature 380*: 499–505.
Dawkins, R. 1986. *The blind watchmaker.* New York: W. W. Norton.
Dawkins, R., and Krebs, J. R. 1979. Arms races between and within species. *Proceedings of the Royal Society of London, B, 205*: 489–511.
DeFelipe, J., and Jones, E. G. 1988. *Cajal on the cerebral cortex.* Oxford: Oxford University Press.
DeLattre, P. 1965. *Comparing the phonetic features of English, French, German, and Spanish.* The Hague: Mouton.

Dennett, D. 1995. *Darwin's dangerous idea.* New York: Simon and Schuster.
Derwing, B., and Baker, W. 1979. Recent research on the acquisition of English morphology. In P. Fletcher and M. Garman, eds., *Language acquisition.* Cambridge: Cambridge University Press.
Deshmukh, M., and Johnson, E. M., Jr. 1997. Programmed cell death in neurons: focus on the pathway of nerve growth factor deprivation–induced death of sympathetic neurons. *Molecular Pharmacology 51* (6): 897–906.
Dickerson, W. B. 1975. Predicting word stress: generative clues in an ESL context. *TESL Studies 1*: 38–52.
Dickerson, W. B., and Finney, R. H. 1978. Spelling in TESL: stress cues to vowel quality. *TESOL Quarterly 12*: 163–76.
Douglas, R. J., and Martin, K. A. C. 1990. Neocortex. In G. M. Shepherd, ed., *The synaptic organization of the brain.* New York: Oxford University Press.
Ebbinghaus, H. [1913] 1964. *Memory: a contribution to experimental psychology.* New York: Dover.
Eccles, J. C. 1977. *The understanding of the brain.* 2d ed. New York: McGraw-Hill.
Edelman, G. 1987. *Neural Darwinism.* New York: Basic Books.
Eden, G. F., Stein, J. F., and Wood, F. B. 1994. Differences in eye movements and reading problems in dyslexic and normal children. *Vision Research 34* (10): 1345–58.
Eimas, P. D., Siqueland, E. R., Jusczyk, P., and Vigorito, J. 1971. Speech perception in infants. *Science 171*: 303–6.
Ellias, S. A., and Grossberg, S. 1975. Pattern formation, contrast control, and oscillations in the short term memory of shunting on-center off-surround networks. *Biological Cybernetics 20*: 69–98.
Errington, M. C., and Bliss, T. V. P. 1982. Long-lasting potentiation of the perforant pathway *in vivo* is associated with increased glutamate release. *Nature 297*: 496–98.
Ferguson, C. A., and Slobin, D. I. 1973. *Studies of child language development.* New York: Holt, Rinehart, and Winston.
Fernald, A., and Kuhl, P. K. 1987. Acoustic determinants of infant preference for motherese speech. *Infant Behavior and Development 10* (3): 279–93.
Fiala, J. C., Grossberg, S., and Bullock, D. 1996. Metabotropic glutamate receptor activation in cerebellar Purkinje cells as substrate for adaptive timing of the classically conditioned eye-blink response. *Journal of Neuroscience 16* (11): 3760–74.
Fillmore, C. 1968. The case for case. In E. Bach and R. Harms, eds., *Universals in linguistic theory.* New York: Holt, Rinehart, and Winston.
Flege, J. E. 1988. The production and perception of foreign language speech sounds. In H. Winitz, ed., *Human communication and its disorders.* Norwood, NJ: Ablex.
Fletcher, P., and Garman, M., eds. 1979. *Language acquisition.* Cambridge: Cambridge University Press.
Fox, C. 1962. *The structure of the cerebellar cortex.* New York: Macmillan.
Fraser, B. 1976. *The verb-particle construction in English.* New York: Academic Press.
Fromkin, V. A., ed. 1973. *Speech errors as linguistic evidence.* The Hague: Mouton.
———. 1977. *Errors in linguistic performance.* New York: Academic Press.
Fromkin, V., and Rodman, R. 1974. *An introduction to language.* New York: Holt, Rinehart, and Winston.
Gardner, R. A., and Gardner, B. T. 1969. Teaching sign language to a chimpanzee. *Science 165*: 664–72.
Garrett, M. 1993. Errors and their relevance for models of language production. In G. Blanken, J. Dittmann, J. Grimm, J. C. Marshall, and C. W. Wallesch, eds., *Linguistic disorders and pathologies.* Berlin: de Gruyter.

Geschwind, N. 1972. Language and the brain. *Scientific American 242* (April): 76–83.
Geschwind, N., and Galaburda, A. M., eds. 1987. *Cerebral lateralization: biological mechanisms, associations, and pathology.* Cambridge: MIT Press.
Geschwind, N., and Levitsky, W. 1968. Human brain: left-right asymmetries in temporal speech region. *Science 161*: 186–187.
Givón, T. 1993. *English grammar: a function-based introduction.* Amsterdam: J. Benjamins.
Gold, E. 1965. Limiting recursion. *Journal of Symbolic Logic 30*: 1–19.
———. 1967. Language identification in the limit. *Information and Control 10*: 447–74.
Goldsmith, J. 1993. *The last phonological rule: reflections on constraints and derivations.* Chicago: University of Chicago Press.
———. 1994. Grammar within a neural network. In S. D. Lima, R. L. Corrigan, and G. K. Iverson, eds., *The reality of linguistic rules.* Amsterdam: John Benjamins.
Gopnik, M. 1990. Genetic basis of grammar defect. *Naure 346*: 226.
Gopnik, M., and Crago, M. B. 1991. Familial aggregation of a developmental language disorder. *Cognition 39*: 1–50.
Granit, R. 1948. The off/on ratio of the isolated on-off elements in the mammalian eye. *British Journal of Ophthalmology 32*: 550–61.
Gray, E. G. 1970. The fine structure of nerve. *Comparative Biochemistry and Physiology 36*: 419.
Gray, J. S. 1975. *Elements of a two-process theory of learning.* London: Academic Press.
———. 1982. *The neuropsychiatry of anxiety: an enquiry into the functions of the septo-hippocampal system.* Oxford: Oxford University Press.
Gray, J. S., and Rawlins, J. N. P. 1986. Comparator and buffer memory: an attempt to integrate two models of hippocampal function. In R. L. Isaacson and K. Pribram, eds., *The hippocampus,* vol. 4. New York: Plenum.
Greenberg, J. 1968. Some universals of grammar. In J. Greenberg, *Universals of Language.* Cambridge: MIT Press.
Grice, H. P. 1975. Logic and conversation. In P. Cole and J. L. Morgan, eds., *Syntax and semantics,* vol. 3, *Speech acts.* New York: Academic Press.
Grossberg, S. 1968. Some physiological and biochemical consequences of psychological postulates. *Proceedings of the National Academy of Sciences 60*: 758–65. Reprinted in Grossberg 1982b.
———. 1972a. Neural expectation: cerebellar and retinal analogs of cells fired by learnable or unlearned pattern classes. *Kybernetik 10*: 49–57.
———. 1972b. A neural theory of punishment and avoidance. I: qualitative theory. *Mathematical Biosciences 15*: 39–67.
———. 1972c. A neural theory of punishment and avoidance. II: Quantitative theory. *Mathematical Biosciences 15*: 253–85. Reprinted in Grossberg, 1982b.
———. 1975. A neural model of attention, reinforcement, and discrimination learning. *International Review of Neurobiology 18*: 263–327.
———. 1980. How does a brain build a cognitive code? *Psychological Review 87*: 1–51. Reprinted in Grossberg 1982b.
———. 1982a. Processing of expected and unexpected events during conditioning and attention: a psychophysiological theory. *Psychological Review 89*: 529–72.
———. 1982b. *Studies of mind and brain: neural principles of learning, perception, development, cognition and motor control.* Dordrecht: Reidel.
———. 1986. The adaptive self-organization of serial order in behavior: speech, language, and motor control. In E. C. Schwab and H. C. Nusbaum, eds., *Pattern recognition by humans and machines.* Vol. 1. Orlando: Academic Press.

Grossberg, S., and Merrill, J. W. L. 1992. A neural network model of adaptively timed reinforcement learning and hippocampal dynamics. *Cognitive Brain Research 3*: 3–38.

Grossberg, S., and Olson, S. J. 1994. Rules for the cortical map of ocular dominance and orientation columns. *Neural Networks 7* (6/7): 883–94.

Grossberg, S., and Stone, G. 1986. Neural dynamics of word recognition and recall: attentional priming, learning, and resonance. *Psychological Review 93* (1): 46–74.

Gruber, J. S. [1965]1976. *Lexical structures in syntax and semantics*. The Hague: Mouton.

Halle, M., Hughes, G. W., and Radley, J.-P. A. 1957. Acoustic properties of stop consonants. *Journal of the Acoustic Society of America 29*: 107–16.

Hamburger, H., and Wexler, K. 1975. A mathematical theory of learning transformational grammar. *Journal of Mathematical Psychology 12*: 137–77.

Hamburger, V. 1975. Cell death in the development of the lateral motor column of the chick embryo. *Journal of Comparative Neurology 160*: 535–46.

Harrison, J. M., and Howe, M. E. 1974. Anatomy of the afferent auditory nervous system of mammals. In W. D. Keidel and W. D. Neff, eds., *Auditory system anatomy and physiology*. Berlin: Springer-Verlag.

Hartline, H. K. 1949. Inhibition of activity of visual receptors by illuminating nearby retinal elements in the *Limulus* eye. *Federal Proceedings 8*: 69–84.

Hartline, H. K., and Graham, C. H. 1932. Nerve impulses from single receptors in the eye. *Journal of Cellular and Comparative Physiology 1*: 277–95.

Hartline, H. K., and Ratliff, F. 1954. Spatial summation of inhibitory influences in the eye of *Limulus*. *Science 120*: 781.

Hattori, T., and Suga, N. 1997. The inferior colliculus of the mustached bat has the frequency-vs.-latency coordinates. *Journal of Comparative Physiology, A, 180* (3): 271–84.

Hebb, D. O. 1949. *The organization of behavior*. New York: Wiley.

Heraclitus. 1987. *Fragments*. Trans. T. M. Robinson. Toronto: University of Toronto Press.

Hinton, G. E. 1977. "Relaxation and its role in vision." Doctoral diss., University of Edinburgh.

Holmes, G. 1939. The cerebellum of man. *Brain 62*: 1–30.

Holmes, J. A., and Singer, H. 1966. *Speed and power of reading in high school*. Washington: U.S. Department of Health, Education, and Welfare, Office of Education.

Hopfield, J. J. 1982. Neural networks and physical systems with emergent collective computational abilities. *Proceedings of the National Academy of Sciences (USA) 79*: 2554–58.

Hubel, D. H., and Livingstone, M. S. 1987. Segregation of form, color, and stereopsis in primate area 18. *Journal of Neuroscience 7*: 3378–415.

Hubel, D. H., and Wiesel, T. N. 1977. Functional architecture of macaque monkey visual cortex. *Proceedings of the Royal Society of London B, 198*: 1–59.

Hurst, J. A., Baraitser, M., Auger, E., Graham, F., and Norell, S. 1990. An extended family with an inherited speech disorder. *Developmental Medicine and Child Neurology 32*: 347–55.

Isaacson, R. L., and Pribram, K., eds. 1975–86. *The hippocampus*. 4 vols. New York: Plenum.

Ivry, R. B., and Keele, S. W. 1989. Timing functions of the cerebellum. *Journal of Cognitive Neuroscience 1*: 136–52.

Jackendoff, R. 1994. *Patterns in the mind*. New York: Basic Books.

Jaeger, J. J., Lockwood, A. H., Kemmerer, D. L., Van-Valin, R. D., Murphy, B. W., and Khalak, H. G. 1996. A positron emission tomographic study of regular and irregular verb morphology in English. *Language 72* (3): 451–97.

Jakobson, R. 1968. *Child language, aphasia, and phonological universals*. The Hague: Mouton.

Jones, E. G. 1981. Anatomy of cerebral cortex: columnar input-output relations. In F. O. Schmidt, F. G. Worden, G. Adelman, and S. G. Dennis, eds., *The cerebral cortex*. Cambridge: MIT Press.
Joseph, R. 1993. *The naked neuron: evolution and the languages of the body and brain*. New York: Plenum Press.
Jusczyk, P., and Hohne, E. A. 1997. Infants' memory for spoken words. *Science 277*: 1984–86.
Kandel, E. R., and Hawkins, R. D. 1992. The biological basis of learning and individuality. *Scientific American 267* (3): 78–87.
Kasamatsu, T. 1983. Neuronal plasticity by the central norepinephrine system in the cat visual cortex. *Progress in Psychobiology and Physiological Psychology 10*: 1–112.
Kasamatsu, T., Pettigrew, J., and Ary, M. 1979. Restoration of visual cortical plasticity by local microperfusion of norepinephrine. *Journal of Comparative Neurology 184*: 163–82.
Kelly, J. P. 1985. Auditory system. In E. R. Kandel and J. H. Schwartz, eds., *Principles of neural science*, 2d ed. New York: Elsevier.
Kent, R. D., Netsell, R., and Abbs, J. H. 1979. Acoustic characteristics of dysarthria associated with cerebellar disease. *Journal of Speech and Hearing Research 22*: 627–48.
Kim, K. H. S., Relkin, N. R., Lee, K-M, and Hirsch, J. 1997. Distinct cortical areas associated with native and second languages. *Nature 388*: 171–74.
Kimura, D. 1967. Functional asymmetry of the brain in dichotic listening. *Cortex 3*: 163–178.
———. 1988. Reveiw of *What the hands reveal about the brain*. *Language and Speech 31* (4): 375–78.
Kirkwood, J. R. 1995. *Essentials of neuroimaging*. New York: Churchill Livingstone.
Klein, D., Zatorre, R. J., Milner, B., Meyer, E., and Evans, A. C. 1994. Left putaminal activation when speaking a second language: evidence from PET. *Neuroreport 5* (17): 2295–97.
Klein, M., and Kandel, E. R. 1978. Presynaptic modulation of voltage-dependent $Ca^{++}$ current: mechanism for behavioral sensitization. *Proceedings of the National Academy of Sciences (USA) 75*: 3512–16.
———. 1980. Mechanism of calcium current modulation underlying presynaptic facilitation and behavioral sensitization in *Aplysia*. *Proceedings of the National Academy of Sciences (USA) 77*: 6912–16.
Klima, E. S., and Bellugi, U. 1979. *The signs of language*. Cambridge: Harvard University Press.
Krashen, S. 1982. *Principles and practice of second language acquisition*. New York: Pergamon.
Kripkee, B., Lynn, R., Madsen, J. A., and Gay, P. E. 1982. Familial learning disability, easy fatigue, and maladroitness: preliminary trial of monosodium glutamate in adults. *Developmental Medicine and Child Neurology 24*: 745–51.
Kuhl, P. K. 1983. Perception of auditory equivalence classes for speech in early infancy. *Infant Behavior and Development 6* (3): 263–85.
———. 1991. Human adults and human infants show a "perceptual magnet effect" for the prototypes of speech categories, monkeys do not. *Perception and Psychophysics 50* (2): 93–107.
Kuhl, P. K., and Miller, J. D. 1975. Speech perception by the chinchilla: voiced-voiceless distinction in alveolar plosive consonants. *Science 190*: 69–72.
Kuhn, T. S. 1957. *The Copernican revolution: planetary astronomy in the development of Western thought*. Cambridge: Harvard University Press.

———. 1962. *The structure of scientific revolutions*. Chicago: University of Chicago Press.
Lahey, M., Liebergott, J., Chesnick, M., Menyuk, P., and Adams, J. 1992. Variability in children's use of grammatical morphemes. *Applied Psycholinguistics 13*: 373–98.
Lashley, K. S. 1950. In search of the engram. In *Society of Experimental Biology Symposium no. 4: Psychological mechanisms in animal behaviour*. London: Cambridge University Press.
———. 1951. The problem of serial order in behavior. In L. A. Jeffress, ed., *Cerebral mechanisms in behavior*. New York: Wiley.
Leckman, J. F., Peterson, B. S., Anderson, G. M., Arnsten, A. F. T., Pauls, D. L., and Cohen, D. J. 1997. Pathogenesis of Tourette's syndrome. *Journal of Child Psychology and Psychiatry 38* (1): 119–42.
Lehmann, W. P. 1978. Conclusion: toward an understanding of the profound unity underlying languages. In W. P. Lehmann, ed., *Syntactic typology: studies in the phenomenology of language*. Austin: University of Texas Press.
Lenneberg, E. H. 1967. *Biological foundations of language*. New York: Wiley.
Leonard, L. B. 1996. Characterizing specific language impairment: a crosslinguistic perspective. In M. L. Rice, ed, *Toward a genetics of language*. Mahwah, NJ: Lawrence Erlbaum.
LeVay, S., Connolly, M., Houde, J., and Van Essen, D. C. 1985. The complete pattern of ocular dominance stripes in the striate cortex and visual field of the macaque monkey. *Journal of Neuroscience 5*: 486–501.
LeVay, S., Wiesel, T. N., and Hubel, D. H. 1980. The development of ocular dominance columns in normal and visually deprived monkeys. *Journal of Comparative Neurology 191*: 1–51.
Li, C. N., ed. 1976. *Subject and topic*. New York: Academic Press.
Li, C. N., and Thompson, S. A. 1981. *Mandarin Chinese: a functional reference grammar*. Berkeley: University of California Press.
Liberman, A. M., DeLattre, P. C., and Cooper, F. S. 1952. The role of selected stimulus variables in the perception of unvoiced stop consonants. *American Journal of Psychology 65*: 497–516.
Liberman, A. M., Harris, K. S., Hoffman, H. S., and Griffith, B. L. 1957. The discrimination of speech events within and across phoneme boundaries. *Journal of Experimental Psychology* 54: 358–68.
Liberman, M. 1979. *The intonational system of English*. New York: Garland.
Lichtheim, L. 1885. On aphasia. *Brain 7*: 443.
Lieberman, P. 1968. Primate vocalizations and human linguistic ability. *Journal of the Acoustical Society of America 44*: 1574–84.
———. 1975. *On the origins of language: an introduction to the evolution of human speech*. New York: Macmillan.
———. 1984. *The biology and evolution of language*. Cambridge: Harvard University Press.
———. 1991. *Uniquely human: the evolution of speech, thought, and selfless behavior*. Cambridge: Harvard University Press.
Lima, S. D., Corrigan, R. L., and Iverson, G. K., eds. 1994. *The reality of linguistic rules*. Amsterdam: John Benjamins.
Lisker, L., and Abramson, A. S. 1964. A cross-language study of voicing in initial stops: acoustical measurements. *Word 20* (3): 384–422.
Livingstone, M. S., and Hubel, D. H. 1984. Anatomy and physiology of a color system in the primate visual cortex. *Journal of Neuroscience 4*: 2830–35.
———. 1987. Psychophysical evidence for separate channels for the perception of form, color, movement, and depth. *Journal of Neuroscience 7*: 3416–68.

Llinás, R. R., and Hillman, D. E. 1969. Physiological and morphological organization of the cerebellar circuits in various vertebrates. R. R. In Llinás, ed., *Neurobiology of cerebellar evolution and development*. Chicago: American Medical Association.

Long, M. H. 1990. Maturational constraints on language development. *Studies in Second Language Acquisition 12*: 251–74.

Lorento de Nó, R. 1943. Cerebral cortex: architecture, intracortical connections, motor projections. In J. F. Fulton, ed., *Physiology of the nervous system*. New York: Oxford University Press.

Loritz, D. 1990. Linguistic hypothesis testing in neural networks. In *Proceedings of the Georgetown University Round Table on Languages and Linguistics*. Washington: Georgetown University Press.

———. 1991. Cerebral and cerebellar models of language learning. *Applied Linguistics 12* (3): 299–318.

Lund, J. S., Boothe, R. G., and Lund, R. D. 1977. Development of neurons in the visual cortex (area 17) of the monkey (*Macaca nemestrina*): a Golgi study from fetal day 27 to postnatal maturity. *Journal of Comparative Neurology 176* (2): 149–88.

Lynch, G. 1986. *Synapses, circuits, and the beginnings of memory*. Cambridge: MIT Press.

Machon, R. A., Mednick, S. A., and Huttunen, M. O. 1997. Adult major affective disorder after prenatal exposure to an influenza epidemic. *Archives of General Psychiatry 54* (4): 322–28.

MacWhinney, B. 1987a. The competition model. In MacWhinney, 1987b.

———, ed. 1987b. *Mechanisms of language acquisition*. Hillsdale, NJ: Lawrence Erlbaum.

MacWhinney, B., and Leinbach, A. J. 1991. Implementations are not conceptualizations: revising the verb learning model. *Cognition 29*: 121–57.

Madison, D. V., and Nicoll, R. A. 1986. Actions of noradrenaline recorded intracellularly in rat hippocampal CA1 pyramidal neurones, *in vitro*. *Journal of Physiology (London) 372*: 221–44.

Mandelbrot, B. 1982. *The fractal geometry of nature*. San Francisco: W. H. Freeman and Co.

Manter, J. T. 1975. *Manter and Gatz's essentials of clinical neuroanatomy and neurophysiology*. 5th ed. (R. G. Clark). Philadelphia: F. A. Davis.

Martin, R. D. 1982. Allometric approaches to the evolution of the primate nervous system. In E. Armstrong and D. Falk, eds., *Primate brain evolution*. New York: Plenum Press.

McClelland, J. L., and Rumelhart, D. E., eds. 1986. *Parallel distributed processing*. Vol. 2. Cambridge: MIT Press.

McCollough, C. 1965. Color adaptation of edge-detectors in the human visual system. *Science 149*: 1115–16.

McCulloch, W. S., and Pitts, W. 1943. A logical calculus of the ideas immanent in nervous activity. *Bulletin of Mathematical Biophysics 5*: 115–33.

McGlade-McCulloh, E., Yamamoto, H., Tan, S.-E., Brickey, D. A., and Soderling, T. R. 1993. Phosphorylation and regulation of glutamate receptors by calcium/calmodulin-dependent protein kinase II. *Nature 362*: 640.

McLuhan, M. 1965. *The Gutenberg galaxy: the making of typographic man*. Buffalo: University of Toronto Press.

McNeill, D. 1966. Developmental psycholinguistics. In F. Smith and G. Miller, eds., *The genesis of language*. Cambridge: MIT Press.

Mednick, S. A. 1994. Prenatal influenza infections and adult schizophrenia. *Schizophrenia Bulletin 20* (2): 263–67.

Menn, L., and Obler, L. 1990. *Agrammatic aphasia: a crosslanguage narrative sourcebook*. Amsterdam: John Benjamins.

Menyuk, P. 1977. *Language and maturation.* Cambridge: MIT Press.
―――. 1988. *Language development: knowledge and use.* Boston: Scott, Foresman/Little, Brown.
Merzenich, M. M., Jenkins, W. M., Johnston, P., Schreiner, C., Miller, S. L., and Tallal, P. 1996. Temporal processing deficits of language-learning impaired children ameliorated by training. *Science 271*: 77–81.
Metter, E. J., Riege, W. H., Hanson, W. R., Camras, L. R., Phelps, M. E., and Kuhl, D. E. 1983. Correlations of glucose metabolism and structural damage to language function in aphasia. *Brain and Language 21* (2): 187–207.
Miller, G. A. 1956. The magic number seven, plus or minus two. *Psychological Review 63*: 81–97.
Miller, G. A., and Gildea, P. M. 1987. How children learn words. *Scientific American 257* (9): 94–99.
Milner, B. 1966. Amnesia following operation on the temporal lobes. In C. W. M. Whitty and O. Zangwill, eds., *Amnesia.* London: Butterworth.
Milner, B., Corkin, S., and Teuber, H.-L. 1968. Further analysis of the hippocampal amnesic syndrome: 14-year follow-up study of H. M. *Neuropsychologia 6*: 215–34.
Minsky, M., and Papert, S. 1967. Perceptrons and pattern recognition. Artificial Intelligence Memo no. 140. MAC-M-358. Project MAC. Cambridge, MA. Sept.
―――. 1969. *Perceptrons.* Cambridge: MIT Press.
Montemurro, D. G., and Bruni, J. E. 1988. *The human brain in dissection.* New York: Oxford University Press.
Morton, J. 1969. Interaction of information in word recognition. *Psychological Review 76*: 165–78.
―――. 1979. Word recognition. In Morton and Marshall, 1979.
Morton, J., and Marshall, J. C. 1979. *Psycholinguistics Series II.* London: Elek Scientific Books.
Mountcastle, V. B. 1957. Modality and topographic properties of single neurons of cat's somatic sensory cortex. *Journal of Neurophysiology 20*: 408–34.
Nelson, K. E., and van Kleeck, A. 1987. *Children's language.* Hillsdale, NJ: Lawrence Erlbaum.
Neville, H., and Bavelier, D. 1996. L'extension des aires visuelles chez les sourds: les cortex visuel et auditif ne sont pas aussi distincts qu'on le croit. *Recherche 289*: 90–93.
Nicolson, R. I., Fawcett, A. J., and Dean, P. 1995. *Proceedings of the Royal Society, B: Biological Sciences 259* (1354): 43–47.
Nietzsche, F. W. (1924 [1833]. *Jenseits von Gut und Böse: Vorspiel einer Philosophie der Zukunft.* Leipzig: A. Kröner.
Nigrin, A. 1993. *Neural networks for pattern recognition.* Cambridge: MIT Press.
Ohlemiller, K. K., Kanwal, J. S., and Suga, N. 1996. Facilitative responses to species-specific calls in cortical FM-FM neurons of the mustached bat. *Neuroreport 7* (11): 1749–55.
Olson, D. R. 1991. *Literacy and orality.* Cambridge: Cambridge University Press.
Ong, W. 1982. *Orality and literacy: the technologizing of the word.* London: Methuen.
Orgel, L. E. 1979. Selection in vitro. *Proceedings of the Royal Society of London, B, 205*: 435–42.
Parnas, D. 1972. On the criteria to be used in decomposing systems into modules. *Communications of the Association for Computing Machinery 15* (12): 1053–58.
Patterson, F. 1978. Conversations with a gorilla. *National Geographic 154*: 438–65.
Peirce, C. S. 1877. The fixation of belief. *Popular Science Monthly 12*: 1–15.

———. 1878. How to make our ideas clear. *Popular Science Monthly* 13 (Jan.): 286–302.
———. 1905. What pragmatism is. *The Monist* 15 (Apr.): 161–81. Reprinted in Wiener 1958.
Penfield, W., and Rasmussen, T. 1950. *The cerebral cortex of man*. New York: Macmillan.
Penfield, W., and Roberts, L. 1959. *Speech and brain mechanisms*. Princeton: Princeton University Press.
Peters, A., and Jones, E. G. 1984. *Cerebral cortex*. Vol. 1, *Cellular components of the cerebral cortex*. New York: Plenum.
Peterson, G. E., and Barney, H. L. 1952. Control methods used in a study of the vowels. *Journal of the Acoustical Society of America* 24: 175–84.
Piaget, J. 1975. *L'équilibration des structures cognitives: problème central du développment*. Paris: Presses universitaires de France.
Pierrehumbert, J. 1987. *The phonology and phonetics of English intonation*. Bloomington: Indiana University Linguistics Club.
Pierson, M., and Snyder-Keller, A. 1994. Development of frequency-selective domains in inferior colliculus of normal and neonatally noise-exposed rats. *Brain Research* 636: 55–67.
Pinker, S. 1984. *Language learnability and language learning*. Cambridge: Harvard University Press.
———. 1989. *Learnability and cognition: the acquisition of argument structure*. Cambridge: MIT Press.
———. 1994. *The language instinct*. New York: Morrow.
Pinker, S., and Bloom, P. 1990. Natural language and natural selection. *Behavioral and brain sciences* 13: 707–84.
Pinker, S., and Prince, A. 1988. On language and connectionism: an analysis of a distributed processing model of language acquisition. *Cognition* 28: 73–194.
———. 1994. Regular and irregular morphology and the psychological status of rules of grammar. In Lima, Corrigan, and Iverson, 1994.
Plunkett, K. 1995. Connectionist approaches to language acquisition. In P. Fletcher and B. MacWhinney, eds., *The Handbook of Child Language*. Oxford: Blackwell.
Poizner, H., Klima, E. S., and Bellugi, U. 1987. *What the hands reveal about the brain*. Cambridge: MIT Press.
Poritsky, R. 1969. Two and three dimensional ultrastructure of boutons and glial cells in the motoneuronal surface of the cat spinal cord. *Journal of Comparative Neurology* 135: 423.
Premack, D. 1985. "Gavagai!" or the future history of the animal language controversy. *Cognition* 19: 207–96.
Pylyshyn, Z. 1979. Complexity and the study of artificial and human intelligence. In M. Ringle, ed., *Philosophical perspectives in artificial intelligence*. Atlantic Highlands, NJ: Humanities Press.
Quillian, M. R. 1968. Semantic memory. In M. Minsky, ed., *Semantic information processing*. Cambridge: MIT Press.
Ramón y Cajal, S. 1911. *Histologie du système nerveux de l'homme et des vertébrés*. Paris: Maloine.
———. 1955. *Histologie du système nerveux*. Madrid: Consejo Superior de Investigationes Cientificas, Instituto Ramón y Cajal.
Ratliff, F. 1965. *Mach bands*. San Francisco: Holden-Day.
Ratner, N. B. 1993. Interactive influences on phonological behavior: a case-study. *Journal of Child Language* 20: 191–97.

Rauschecker, J. P., Tian, B., and Hauser, M. 1995. Processing of complex sounds in the macaque nonprimary auditory cortex. *Science 268*: 111–14.

Raymond, J. E., Ogden, N. A., Fagan, J. E., and Kaplan, B. J. 1988. Fixational instability and saccadic eye movements of dyslexic children with subtle cerebellar dysfunction. *American Journal of Optometry and Physiological Optics 65* (3): 174–81.

Reilly, J. S., Klima, E. S., and Bellugi, U. 1991. Once more with feeling: affect and language in atypical populations. *Developmental Psychopathology 2*: 367–91.

Rice, M. L., ed. 1996. *Toward a genetics of language.* Mahwah, NJ: Lawrence Erlbaum.

Rice, M. L., and Wexler, K. 1996. Toward tense as a clinical marker of specific language impairment in English-speaking children. *Journal of Speech and Hearing Research 39*: 1239–57.

Rochet, B. L. 1991. Perception ot the high vowel continuum: a crosslanguage study. Paper presented at the International Congress on Phonetic Sciences, Aix-en-Provence.

Roe, A. W., Pallas, S. L., Hahm, J.-O. and Sur, M. 1990. A map of visual space induced in primary auditory cortex. *Science 250*: 818–20.

Rose, W. J. 1993. "Computational adaptation, real-time organization and language learning (CAROLL)." Ph.D. diss. Georgetown University.

Rosenblatt, F. 1958. The perceptron: a probabilistic model for information storage and organization in the brain. *Psychological Review 65*: 386–408.

———. 1959. Two theorems of statistical separability in the perceptron. In *Mechanisation of thought processes*, vol. 1. Proceedings of a Symposium Held at the National Physics Laboratory, 1958. London: H. M. Stationery Office.

———. 1961. *Principles of neurodynamics: perceptrons and the theory of brain mechanisms.* Washington: Spartan Books.

Ross, C. 1978. The rightmost principle of sentence negation. In *Papers from the Fourteenth Regular Meeting of the Chicago Linguistics Society.* Chicago: Chicago Linguistics Society.

Roszak, T. 1986. *The cult of information: the folklore of computers and the true art of thinking.* New York: Pantheon.

Rumelhart, D., Hinton, G., and Williams, R.. 1986. Learning internal representations by error propagation. In Rumelhart and McClelland 1986b.

Rumelhart, D. E., and McClelland, J. L., eds. 1986a. On learning the past tenses of English verbs. In McClelland and Rumelhart 1986.

———, eds. 1986b. *Parallel distributed processing.* Vol. 1. Cambridge: MIT Press.

Sachs, J., Bard, B., and Johnson, M. L. 1981. Language learning with restricted input: case studies of two hearing children of deaf parents. *Applied Psycholinguistics 2* (1): 33–54.

Sadler, L. S., Robinson, L. K., Verdaasdonk, K. R., and Gingell, R. 1994. The Williams syndrome: evidence for possible autosomal dominant inheritance. *American Journal of Medical Genetics 47:* 468–70.

Sawusch, J. R., and Nusbaum, H. C. 1979. Contextual effects in vowel perception. I: Anchor-induced contrast effects. *Perception and Psychophysics 27*: 421–34.

Sawusch, J. R., Nusbaum, H. C., and Schwab, E. C. 1980. Contextual effects in vowel perception. II: Evidence for two processing mechanisms. *Perception and Psychophysics 27*: 421–34.

Schmidt, F. O., Worden, F. G., Adelman G., and Dennis, S. G., eds. 1981. *The cerebral cortex.* Cambridge: MIT Press.

Schmidt, R. 1993. Awareness and second language acquisition. *Annual Review of Applied Linguistics 13*: 206–26.

———. 1994. Implicit learning and the cognitive unconscious: of artificial grammars and SLA. In N. Ellis, ed., *Implicit and explicit learning of languages*. New York: Academic Press.

Schoenle, P. W., and Groene, B. 1993. Cerebellar dysarthria. In Blanken, Dittmann, Grimm, Marshall, and Wallesch, 1993.

Schwab, E. C., and Nusbaum, H. C., eds. 1986. *Pattern recognition by humans and machines*. Vol. 1, *Speech perception*. Orlando: Academic Press.

Selinker, L. 1972. Interlanguage. *International Review of Applied Linguistics 10*: 209–31.

Selkirk, E. 1984. *Phonology and syntax: the relation between sound and structure*. Cambridge: MIT Press.

Shepherd, G. M., ed. 1997. *The synaptic organization of the brain*. Fourth edition. New York: Oxford University Press.

Sherrington, C. S. [1906] 1961. *The integrative action of the nervous system*. New Haven: Yale University Press.

Simon, H. A., and Feigenbaum, E. A. 1979. A theory of the serial position effect. In H. A. Simon, ed., *Models of thought*. New Haven: Yale University Press.

Skinner, B. F. 1957. *Verbal behavior*. New York: Appleton-Century-Crofts.

———. 1971. *Beyond freedom and dignity*. New York: Knopf.

Skrede, K. K., and Malthe-Sorenssen, D. 1981. Increased resting and evoked release of transmitter following repetitive electrical tetanization in hippocampus: a biochemical correlate to long-lasting synaptic potentiation. *Brain Research 708*: 436–41.

Slobin, D. I. 1973. Cognitive prerequisites for the development of grammar. In Ferguson and Slobin 1973.

Small, S. L., Cottrell, G. W., and Tanenhaus, M. K. 1988. *Lexical ambiguity resolution: perspectives from psycholinguistics, neuropsychology, and artificial intelligence*. Los Altos: Morgan Kaufmann Publishers.

Snyder, S. H., and Bredt, D. S. 1992. Biological roles of nitric oxide. *Scientific American 267* (5): 68–71.

Spencer, H. [1862] 1912. *First principles*. New York: D. Appleton and Co.

Sperry, R. W. 1964. The great cerebral commissure. *Scientific American 210*: 42–52.

———. 1970a. Cerebral dominance in perception. In F. A. Young and D. B. Lindsley, eds., *Early experience in visual information processing in perceptual and reading disorders*. Washington: National Academy of Science.

———. 1970b. Perception in the absence of the neocortical commissures. In *Perception and its disorders*. Research Publication 48. Chicago: Association for Research in Nervous and Mental Disease.

———. 1967. Some effects of disconnecting the cerebral hemispheres. In C. Millikan and F. Darley, eds., *Brain mechanisms underlying speech and language*. New York: Grune and Stratton.

Starbuck, V. N. 1993. "The N400 in recovery from aphasia." Ph.D. diss. Washington: Georgetown University.

Stevens, K. N. 1972. The quantal nature of speech: evidence from articulatory-acoustic data. In E. E. David, Jr., and P. B. Denes, eds., *Human communication: a unified view*. New York: McGraw-Hill.

Suga, N. 1990. Biosonar and neural computation in bats. *Scientific American 262* (June): 60–68.

Suga, N., Zhang, Y., and Yan, J. 1997. Sharpening of frequency tuning by inhibition in the thalamic auditory nucleus of the mustached bat. *Journal of Neurophysiology 77* (4): 2098–114.

Swinney, D. A. 1982. The structure and time-course of information interaction during speech comprehension: lexical segmentation, access, and interpretation. In J. Mehler, E. C. T. Walker, and M. Garrett, eds., *Perspectives on mental representation*. Hillsdale, NJ: Lawrence Erlbaum.
Szentágothai, J. 1969. Architecture of the cerebral cortex. In H. H. Jasper, A. Ward, and A. Pope, eds., *Basic mechanisms of the epilepsies*. Boston: Little, Brown.
Tallal, P., Miller, S. L., Bedi, G., Byma, G., Wang, X., Nagarajan, S. S., Schreiner, C., Jenkins, W. M., and Merzenich, M. 1996. Language comprehension in language-learning impaired children improved with acoustically modified speech. *Science* 271: 81–84.
Tallal, P., and Piercy, M. 1973a. Defects of non-verbal auditory perception in children with developmental aphasia. *Nature* 241: 468–69.
———. 1973b. Developmental aphasia: impaired rate of non-verbal processing as a function of sensory modality. *Neuropsychologia* 11 (4): 389–98.
Tannen, D. 1982. *Spoken and written language: exploring orality and literacy*. Norwood, NJ: ABLEX.
Tesniere, L. [1959] 1969. *Elements de syntaxe structurale*. 2d ed. Paris: Klincksieck.
Thoenen, Hans. 1995. Neurotrophins and neuronal plasticity. *Science* 270: 593–98.
Tolman, E. C. 1932. *Purposive behavior in animals and men*. New York: Century Co.
Turing, A. 1936. On computable numbers with an application to the entscheidung-problem. *Proceedings of the London Mathematics Society* 42: 230–65.
Ullman, M., Corkin, S., Coppola, M., Hickok, G., Growdon, J. H., Koroshetz, W. J., and Pinker, S. 1997. A neural dissociation within language: evidence that the mental dictionary is part of declarative memory, and that grammatical rules are processed by the procedural system. *Journal of Cognitive Neuroscience* 9 (2): 289–99.
Ullman, M., and Gopnik, M. 1994. The production of inflectional morphology in hereditary specific language impairment. *McGill Working Papers in Linguistics* 10 (1–2): 81–118.
Vellutino, F. 1987. Dyslexia. *Scientific American* 256 (3): 34–41.
von Baer, K. E. 1828. *Über die Entwicklungsgeschichte der Thiere; Beobachtung und Reflexion*. Königsberg: Barntiäger.
von der Malsburg, C. 1973. Self-organization of orientation sensitive cells in the striate cortex. *Kybernetik* 14: 85–100.
Von Neumann, J. 1958. *The computer and the brain*. New Haven: Yale University Press.
Wang, P. P., and Bellugi, U. 1993. Williams syndrome, Down syndrome, and cognitive neuroscience. *American Journal of Diseases of Children* 147: 1246–51.
Watson, J. D., and Crick, F. H. C. 1953. A structure for deoxyribose nucleic acid. *Nature* 171: 737.
Welker, E., Armstrong-James, M., Bronchti, G., Ourednik, W., Gheorghita-Baechler, F., Duybois, R., Guernsey, D. L., van der Loos, H., and Neumann, P. E. 1996. Altered sensor processing in the somatosensory cortex of the mouse mutant barrelless. *Science* 271: 1864–67.
Wernicke, C. 1874. *Der aphasische Symptomencomplex*. Breslau: Cohn und Weigert.
West, J. R, Hodges, C. A., and Black, A. C., Jr. 1981. Prenatal exposure to ethanol alters the organization of hippocampal mossy fibers in rats. *Science* 211: 957–59.
Wexler, K., and Culicover, P. 1980. *Formal principles of language acquisition*. Cambridge: MIT Press.
White, E. L. 1989. *Cortical circuits: synaptic organization of the cerebral cortex—structure, function, and theory*. Boston: Birkhaeuser.

Wiener, P. 1958. *Charles S. Peirce: values in a universe of chance.* New York: Doubleday.
Wiesel, T. N., and Hubel, D. H. 1965. Comparison of the effects of unilateral and bilateral eye closure on cortical unit responses in kittens. *Journal of Neurophysiology 218*: 1029–40.
Wiesel, T. N., Hubel, D. H., and Lam, D. M. K. 1974. Autoradiographic demonstration of ocular-dominance columns in the monkey striate cortex by means of transneuronal transport. *Brain Research 79*: 273–79.
Wirth, N. 1971. Program development by stepwise refinement. *Communications of the Association for Computing Machinery 14* (4): 221–27.
Wolf, S. S., Jones, D. W., Knable, M. B., Gorey, J. G., Lee, K. S., Hyde, T. M., Coppola, R., and Weinberger, D. R. 1996. Tourette syndrome: prediction of phenotypic variation in monozygotic twins by caudate nucleus D2 receptor binding. *Science 273*: 1225–27.
Wu, C. P. 1992. "Semantic-based synthesis of Chinese idioms (Chengyu)." Ph.D. diss. Georgetown University.
Yan, J., and Suga, N. 1996. Corticofugal modulation of time-domain processing of biosonar information in bats. *Science 273* (5278): 1100.
Zhang, Y., Suga, N., and Yan, J. 1997. Corticofugal modulation of frequency processing in bat auditory system. *Nature 26*, 387 (6636): 900–903.

# INDEX

abduction (Peircean), 10, 15, 89
accommodation, Piagetian, 87
acetic acid, 23, 24
acetylcholine, 46, 47, 49
Ackley, D., 169
action potential, 40–44, 41 f.
adaptive resonance theory (ART), 74–78
adrenaline, 49, 178
adrenocorticotrophic hormone (ACTH), 178
Adrian, E. D., 7
afferent signals, 53
affine transformation. *See* fractal
agonist-antagonist competition, 78
agreement marking, 149
alcohol, 176
alexia, 65
American sign language (ASL), 22, 67, 175, 195 n. 2.1
amino acids, 23
amnesia, anterograde, 55, 88
amplitude, 91
amygdala, 54
anatomies
    avalanche, 32 f.
    columnar, 70, 71 f.
    minimal, 17, 32, 34
    off-center, off-surround, 62, 63 f.
    on-center, off-surround, 54, 63, 64 f., 73, 75 f., 76, 106, 115, 124,
    planar, 72
    polypole, 114–5, 120–2
    radial, 33 f.
    serial, 31
    tonotopic, 104–6, 115, 120–2
    tonotopic and on-center off-surround, 121–2
AND (logical operation), 81
angular gyrus, 65, 67
animal language, 13, 22
    ape, 101
animals, multicellular, 27
aphasia, 7
    aphasia, childhood, 67, 68
    aphasia, conduction, 66, 67
    Broca's, 127, 177
    motor, 61
apoptosis, 182. *See also* neuroptosis
Archaea, 28
arcuate fasciculus, 38, 66, 67 f., 69, 181, 183, 187
Aristarchus, 21
Aristotle, 4, 164
ART. *See* adaptive resonance theory
articulatory posture, 120
artificial intelligence, 20, 144
Aryan language, 21, 164
Asanuma, C., 49
Asimov, I., 195 n. 2.2
ASL. *See* American sign language

association cortex. *See* lobe, parietal cerebral
atrophy, neural, 70
attention, 130
attentional deficit disorder (ADD), 179
auditory nerve, 62, 69
auditory pathways, 102, 104 f.
augmented transition network (ATN), 167
autoreceptors, 196 n. 3.5
autosomal dominance, 181
axon, 27, 36
  axon, squid giant, 39, 42
  collaterals, 32, 37, 38, 44
  terminal, 42, 87

babbling, 183
back-propagation, 169
Bacon, F., 6
Baker, C. L., 141
baking soda, 23
Bar-Hillel, Y., 166, 167
Barney, H. L., 120
barrels, neural, 71, 75
basal ganglia, 51, 54, 120
basilar membrane, 102
basket cells, 48, 63, 127
behaviorism, 8, 19, 124, 147, 185, 186, 192
Békésy, G. von, 102
Bellugi, U., 176
Berko, J., 141
Bickerton, D., 148
bilingual code-switching, 118
bilingualism, 116, 117, 139, 191, 198 n. 7.2-3
black box, 165
Bliss, T. V. P., 47
Bobrow, D., 167
Bohr, N., 195 n. 1.5
Boltzmann machine, 169
brain modules, 64
brain sections
  horizontal, 5 f.
  lateral, 59 f.
  medial, 53, 59 (figs.)
brain stem, 53
brains
  six-celled, 34, 35 f.
  two-celled, 33

Braine, M., 185, 201 n. 12.16
Broca, P., 7, 58, 173
Broca's area, 60, 61, 66, 174
Brodmann areas, 58, 59 f.
Brown, R., 132, 134, 180
bullet, conical, 6
burst, plosive consonant, 100

calcium ($Ca^{2+}$), 51
calmodulin, 48
CAM kinase II, 48
Cambrian period, 28
Carpenter, G. A., 197 n. 5.5
case grammar, 151, 167
case marking, 149
catecholamine, 50
caudate nucleus, 54, 177
cell, eukaryotic, 24, 26 f.
cell nucleus, 26
cerebellar cortex, 56 f., 57 f.
cerebellar feedback, 129 f.
cerebellar nuclei, 56
cerebellar overshoots, 178
cerebellum, 34, 56, 57, 127, 130, 171, 172, 192, 201 n. 12.7
cerebral asymmetry, 174 f., 200 n. 12.7
cerebrum, 36, 57, 172, 196 n. 4.1
  surgical view, 58
chains, syntactic 147
Chambers, R., 21
chandelier cells, 48, 49 f., 71, 197
Changeux, J.-P., 172
Chao, Y.-R., 150
chinchilla, VOT perception by, 112
chloride, 48
Chomsky, N., 9, 19, 22, 23, 84, 112, 123, 132, 143, 147, 166-7
chordates, 34, 52
chunking. *See* unitization
Church, A., 19, 144
cingulate gyrus, 55
cingulum, 55
circular reaction, 183
Cl⁻. *See* chloride
climbing fibers, 56
clitics, 157
CNV (contingent negative variation), 88
cochlea, 102
cochlear nucleus, 102, 104, 105
coelenterates, 27, 30, 32, 34

Cohen, M. A., 127
Cohen, S., 70
columnar organization. *See* anatomies, columnar
combinations, formula for, 15
commissurectomy, cerebral, 65
comparator theory of hippocampus, 88
complementation of storage sites, 83
computational linguistics, 166
computer games as therapy, 179, 180
computer processing, parallel, 14, 86
computer processing, serial, 14
connectionism, 19
consonants, 99
   alveolar, 100, 101
   fricative, 101
   labial, 100
   nasal, 97
   plosive, 100
   prevoiced, 114
   retroflex, 97
   stop, 101
   unvoiced, 109
   velar, 100, 101
   voiced, 109
consequences, 161
contralateral connection. *See* ipsilateral v. contralateral connection
contrast enhancement, 82, 115
   auditory, 106
contrastive analysis, 117, 165
Cook, W. A., 151
Copernicus, N., 21, 89, 163, 164
corpus callosum, 65, 69
cortex, 196 n. 4.1. *See also* cerebrum, neocortex
Crago, M. B., 177
Crick, F., 24, 49
critical periods, 84, 173, 186
   environmentally induced, 175
cryptography, 165

Dale, H. H., 7
Damasio, H., 174, 196 n. 4.3
damping, 93
Darwin, C., 21
deaf children, 186
death, 14, 23
deep structure, 151, 166
Dennett, D., 170

dentate nucleus of the cerebellum, 129
deperseveration, cerebellar, 136, 138
depolarization, 39, 43
Derwing, B., 141
Descartes, R. 4
dichotic listening, 65
diencephalon, 52, 54
dipoles
   bilingual, 118
   gated, 76, 86, 112, 136, 156
   rhythmic, 136, 138
DNA, 24, 25 f.
dopamine, 47, 50, 53, 54, 177
dysarthria, 66, 127
   ataxic, 131
dyslexia, 178, 180, 190, 200 n. 12.7
dysphasia, 177, 180, 200 n. 12.6

ear, inner, 102
ear, middle, 102
ear, outer, 102
Eden, G. F., 181
edge detection, 83
   auditory, 107
EEG, 195 n. 2.1
efferent signals, 53
Eimas, P. 111, 114
Einstein, A., 20, 163, 195 n. 1.5
elephants, 52
embedded clauses, rhythmic generation of, 160 f.
enigma machine, 165
epilepsy, *grand mal*, 65
epilepsy, temporal lobe, 55
epileptic convulsion, 34
Erlanger, J., 87
evoked potentials, 47, 87, 89
evolution, 21
expansions. *See* recasts
expectancies, contextual, 87

$f_0$. *See* fundamental frequency
fallibility, 163
fascicles, 38
feature filling, 115, 116
feedback, 73, 112
feet, metrical, 134, 139, 184
   planning, 135
Feigenbaum, E. A., 195 n. 1.4
fetal alcohol syndrome (FAS), 176

Fillmore, C., 151
filter, acoustic, 95
filter, vocal tract, 96
flagellates. *See* Mastigophora
fMRI. *See* functional magnetic resonance imagery
formant transitions, 100, 179
formants, children's, 98
formants, female, 98
formants, vowel, 97
fornix, 55
fossilization, 192
fractals, 28
Fraser, B., 167, 199 n. 10.8
free will, 20
frequency domain, 94
functional magnetic resonance imagery (fMRI), 68, 195 n. 2.1
functionalism, 13, 150
functors. *See* morphemes, grammatical
fundamental frequency ($f_0$), 91, 98
   children's, 98
   female, 98

GABA (gamma-amino-butyric acid), 47, 48, 79
Galileo, 164
gamma-amino-butyric acid. *See* GABA
Gasser, H., 87
gates, potassium, 39
gates, sodium, 39
generative theory, 9, 166
Genie, 186
Geschwind, N., 173, 175
Gilman, A. G., 47
given information, 150
globus pallidus, 54
glottal pulse, 93, 94 f.
glottis, 90
glutamate, 47, 179, 178
glutaminergic arousal, 180
Golgi, C., 7, 45
Gopnik, M., 177, 180, 181
granule cells, 56
Gray, J. S., 88
Greenberg, J., 148
Grice, H. P., 150
Grimm, J., 164
Grimm, W., 164

Grossberg, S., 16, 72, 78, 86–88, 127, 169, 179, 183, 197 n. 5.5–6
Gruber, J. S., 151

habituation, 81, 82, 111. *See also* neurotransmitter, depletion of
hair cell, 102, 104
Halle, M., 132
harmonics, 91–3
Harvey, W., 4
hearing, 102
Hebb, D., 45–46
Heisenberg, W., 20
hemisphere, cerebral, 58
hemisphere, left cerebral, 58, 64
hemisphere, right cerebral, 58, 67
Heschl's gyrus. *See* neocortex, primary auditory
Hinton, G. E., 169
hippocampus, 49, 55, 88, 89
hippocampus as comparator, 55
HM, 55, 88, 196 n. 4.3
Hodgkin, A. L., 39
Hohne, E. A., 182
Holmes, J. A., 190
Holy Roman Empire, 164
homunculus, motor, 59, 60 f.
homunculus, somatosensory, 61 f.
Hopfield, J. A., 169
horizontal cells, 182
hormones as neurotransmitters, 54
horseshoe crab. *See* Limulus polyphemus
Hubel, D. H., 71, 84, 173, 182
Huxley, A. F., 39
hyperactivity. *See* attentional deficit disorder
hyperpolarization, 48
hypothalamus, 54–55

Illiac IV, 86, 169
imagination, 193
imitation, 11, 184–6
immediate memory span. *See* memory span, transient
inferior colliculus, 53, 102, 106, 113
inferior olivary complex (inferior olive), 127, 129–30
inhibition
   in cerebellum, 57
   lateral, 34, 75–76, 82, 113

inhibitory surround, radius, 86, 127
innateness of language, 9, 186
innate rules, 146
interference, 116–7, 165, 191
International Phonetic Alphabet (IPA), 95, 109
ipsilateral v. contralateral connection, 65
Isaacson, R. L., 55

Jackendoff, R., 10–11, 15
Jakobson, R., 136
James, W., 162–3
Jones, W., 21, 164
Jusczyk, P., 182

K+. *See* potassium
Kasamatsu, T., 173, 178, 180
Kepler, J., 89
Kim, K. H. S., 174, 200 n. 12.1
Klein, D., 177
knob, presynaptic. *See* axon terminal
koniocortex. *See* neocortex, primary auditory
Kripkee, B., 179–81
Kuhl, P. K., 112, 120
Kuhn, T. S., 163

Lam, D. M. K., 84
Lamarck, J., 21
lambda calculus, 19, 144
laminae, cerebral, 37 f., 68
language
  "acquisition," 9, 171
  acquisition device, 13, 171
  learning, 171, 181
  teaching, 164–5
language-learning impaired (LLI), 179
larynx, 90, 91 f.
Lashley, K., 8, 134–5, 143
lateral geniculate nucleus (LGN), 54, 62, 77
lateral inhibition, 115, 124
lateral lemniscus, 105
lateralization, 58, 64, 173–5
learnability theory, 11, 186
learning, 46, 80
  adult, 191
  backward, 125
  bowed serial, 126 f.

cerebellar, 190
disabilities, 20
opportunistic, 86
pattern, 116
serial, 123–5, 129, 131
Leeuwenhoek, A., 7
left-handedness, 67
Lehmann, W. P., 148
Lenneberg, E. H., 84, 173
Leonard, L. B., 178
Levi-Montalcini, R., 70
Levitsky, W., 173
LGN. *See* lateral geniculate nucleus
Li, C. N., 150
Lichtheim, L., 66
Lieberman, P., 22
limbic system, 53–54, 88
Limulus polyphemus, 62
LISP computer language, 147
lobe
  frontal cerebral, 58
  occipital cerebral, 5 f., 62
  parietal cerebral, 62
  temporal cerebral, 62
Locke, J., 6, 172
locus coeruleus, 53
Loewi, O., 7, 46, 49
logogen model, 168
Lømo, T., 47
long-distance dependencies, 167
long-term memory (LTM), 46–48, 89
  equation for, 79
  invariance of, 83, 118
long-term potentiation (LTP), 47, 48 f.
Loritz, D., 118
loudness, 91
LTM. *See* long-term memory
Lund, J. S., 182
Luther, M., 164

machine translation, 166
Madison, D. V., 50
magic number, 127
magnesium, 48
magnetoencephalogram (MEG), 89
mammilary body, 55
manner, articulatory 101
Mastigophora, 27 f., 28, 36
McCollough effect, 76 f., 197 n. 5.1
McCulloch, W. S., 45

McNeill, D., 11, 132–4, 185
meaning, 161–3, 170
meatus. *See* ear, outer
medial geniculate nucleus (MGN), 54, 102, 105–6, 113
medulla, 53
membrane
  active, 27
  cell, 26
  neural, 38
  nuclear, 26
  passive, 26
memory, complementation of, 118
memory span
  short term, 127
  transient, 127
messengers, reverse, 51
messengers, second, 47
metathesis, 19, 134–9, 143, 181
metrical feet, 181. *See also* feet, metrical
metrical phonology, 134
$Mg^{2+}$. *See* magnesium
MGN. *See* medial geniculate nucleus
midbrain, 53
Miller, G. A., 127
Miller, J. D., 112
Miller, S., 23–24, 195 n. 2.2
Milner, B., 55
Minié, C., 6
Minsky, M., 86, 169, 197 n. 5.7
monosodium glutamate (MSG), 178–80
morphology, 177, 180
  derivational, 140–1
  English plural, 141
  grammatical, 139, 140
  inflectional, 140, 142
  offbeat, 139–40, 155, 178
Morton, J., 168
mossy fibers, 56
motherese, 183
Mountcastle, V., 71
movement, 143
  dative, 154
  particle, 155
  passive, 11, 143
multiple sclerosis, 42
myelin, 42 f., 43 f.
myelin and critical periods, 201 n. 12.10
myelin, 41, 69, 182

N400, 88
$Na^+$. *See* sodium
negation, 158–9, 187–8, 193
neocortex, 196 n. 4.1. *See also* cerebrum
  auditory, 62
  motor, 58
  primary auditory, 62, 68, 69, 102, 106, 115
  primary sensory, 61
  primary visual, 62, 68–69, 73
  sensory, 61
nerve growth factor (NGF), 51, 70
neural Darwinism, 70
neural networks, artificial, 12. *See also* connectionism
neurogenesis, 68
neuron, 27, 36
neuroptosis, 182
neurotransmitter, 45, 46
  depletion of, 81, 87
  retrograde, 51
new information, 150, 188–9
Nicoll, R., 50
Nietzsche, F. W., 5, 8
nitric oxide, 51, 70
NMDA (N-methyl D-aspartate), 47, 79
Nobel prizes in brain science, 14
noise suppression, 82
  auditory, 107
nonlinearity, 80
nonspecific arousal (NSA), 78, 88–89, 117, 136, 159
noradrenaline, 49–50, 53, 173, 178–80, 196 n. 4.4
normalization, 83
normalization, phonemic, 117, 120
notochord, 52
NSA. *See* nonspecific arousal

octopus cells, 104, 105 f., 179
ocular dominance columns, 84
off-center off-surround anatomy. *See* anatomies, off-center, off-surround
old information, 150
oligodendrocyte, 41, 43 f.
Olson, S. J., 197 n. 5.6
ontogeny of language, 172
ontogeny recapitulates phylogeny, 30, 31 f.
outstar, 33 f.
oval window, 102

P300, 88
paleocortex, 54
Papert, 86, 169, 197 n. 5.7
parahippocampal lobe, 55
parallel fiber, 56
Parkinson's disease, 47, 50, 54
passive voice, 143, 152–3
Peirce, C. S., 8, 10, 59, 87, 89, 153, 161–3
Penfield, W., 59
perception
  of sound, 106
  bilingual, 116
  categorical, 109, 111, 115
  phonemic, 109, 115, 116 f.
  subliminal, 106
perceptrons, 86, 169
perseveration, 127. *See also*
  deperseveration
PET. *See* positron-emission-tomography
Peterson, G. E., 120
phone, 95
phoneme, 95
phonics, 191
phospholipid hydrocarbon, 24
phosphorylation, 48
phrase planning, 135
Piaget, J., 87, 183
Pitts, W., 45
pituitary gland, 54
place, 100
plan, 184
planar organization. *See* anatomies,
  planar
planum temporale, 173, 181
plasticity, 68, 173
Plato, 164
polypole. *See* anatomies, polypole
pons, 53
pontine nucleus, 127
positron emission tomography (PET), 89
potassium, 39–41
power spectrum, 94 f.
pragmatics, 150
pragmatism, 162–3
Pribram, K., 55
primacy effect, 125, 149, 188
priming effects, 156, 168
parity, 86
programmed cell death. *See* apoptosis
pronouns, 157

proprioception, 129
Proto-Indo-European, 21
Prozac, 47
puberty, neural, 72
Purkinje cell, 56–57, 129–30
pushdown-store automaton, 19
putamen, 54, 118, 120, 177
pyramidal cells, 36–37, 47–48, 69–70, 75
pyramidal tract, 53, 196 n. 4.2

quantal articulation. *See* speech
  articulation, optimal
quenching threshold (QT), 179
questions, 159–60
Quillian, M. R., 167–8

Ramón y Cajal, S., 7, 15, 45, 57, 182, 195
  n. 1.2
Ranvier, nodes of, 41
raphe nucleus, 53
Rauschecker, J. P., 120
reading, 190
  and Chinese, 201 n.12.17
rebounds, 76, 77 f., 78, 83, 86–87, 136–8,
  156
  vocalic, 117 f.
recasts, 186
recency effect, 125
receptor, neurotransmitter, 46
recursion, 19–20, 144
refractory time, 40
relation gradient, 155, 177
relative clauses, 145, 159
Renshaw interneurons, 40
resonance
  adaptive, 34, 80, 89
  acoustic, 93
reticular formation, brain stem, 49
reticular formation, brain stem, 54
reticular formation (nucleus), thalamic,
  54, 63, 69
retinal afterimage, 76
retrograde messenger, 70
rhythm, 34, 131–2, 141, 172, 180, 190. *See
  also* dipoles, rhythmic
  circadian, 131
  generators, 131, 136
Rice, M. L., 178
Ritalin, 47, 180
Rodbell, M., 47

Rolando, fissure of, 58
*Romeo and Juliet*, 135
Rosenblatt, F., 86, 169
Ross, C., 158, 188
Rumelhart, D., 169

saccades, 178–80
saltatory conduction, 42
Sanskrit, 21
Schwann cell, 41, 43 f.
Sejnowski, T., 169
self-organization, 173
self-similarity
self-similarity, 23–25, 29, 170
semantic features, 167
semantic network, 167–8
semantic wall, 166
semantics, 12
semivowel, 99
sentential rhythm dipole, 159
serial behavior, 125, 134
serial order, 124
serial order gradient, 127
serial processing, 147. *See also* computer processing, serial
serial theory, 123
serotonin, 47, 53
sex, 30
Shepherd, G. M., 196 n. 4.4
Sherrington, C. S., 7, 45, 78
shunting equations, 80
sigmoidal signal function, 80
sign language. *See* American sign language
Simon, H. A., 195 n. 4
Singer, H., 190
Skinner, B. F., 8, 123, 184
social Darwinism, 8
sociolinguistics, 150
sodium, 39–41
sound, physics of, 90
sound localization, 105–6
sound pressure of speech, 102
specific language impairments (SLI), 180
speech, 90
speech articulation, 100
  optimal, 102 f.
speech sounds, periodic, 90
speech sounds, aperiodic, 99
Spencer, H., 8

Sperry, 65
spike. *See* action potential
spine, dendritic, 42
split-brain. *See* commisurectomy, cerebral
Spooner, W. A., 19, 134
Spooner circuit, 138 f.
spoonerism, foot-timed, 135
spoonerism, syllable-timed, 136
spoonerisms, 19, 134, 139
Spooner's circuit, 135, 136 f., 138 f.
spreading activation models, 168
stellate cells, 71
stimulus-response chains, 123–4, 147
stress
  contrastive, 154
  English, 132
  French, 198 n. 9.2
  patterns, 134
  related effects in Chinese, 198 n. 9.2
  Russian, 199 n. 9.3
  Turkish, 199 n. 9.3
stress-timed languages, 139
striate cortex. *See* neocortex, primary visual
stuttering, 130
subject selection hierarchy, 152
subject, 148–9
substantia nigra, 53
subtopics. *See* topics and subtopics
Suga, N., 114
superior colliculus, 53
superior olivary complex (superior olive), 105
syllable planning, 135
symmetry, bilateral, 34
synapse, 44 f., 46
  gap junction, 44, 45
syntax, 188
  rule-governed, 9, 12
Szenthágothai, J., 71, 197 n. 4.6

tabula rasa, 6, 115, 172, 192
Tallal, P., 179–80
tempo, 130–1
temporal processing deficits, 179
teratogens, 176, 200 n. 12.2
terrible twos, 187
Tesniere, L., 151

testosterone, 174
thalamus, 54, 129
    reticular formation (nucleus), 54, 63, 69
thematic relations, 151
thermophilic life. *See* Archaea
Thompson, S., 150
thresholds, 79, 124
    neuron, 42
    quenching, 80
time domain, 95
tip-of-the-tongue phenomenon (TOT), 132–4, 184, 190–1
Tolman, E. C., 87
tonotopic organization. *See* anatomies, tonotopic
topic gradient, 153–4
topics and subtopics, 149–51, 154, 186
topicalization, 152, 188
topic-comment grammar, 150
TOT. *See* tip-of-the-tongue phenomenon
Tourette syndrome, 177, 181
Tower of Hanoi problem, 55, 88
traces, syntactic 158
transformations. *See* movement
trees, binary, 19
truth, 161–3
Turing machine, 166
Turing, A., 13, 144

unitization, 127, 132, 138
universal order, 148
    subject-verb, 149
    topic-verb, 149

ventricles, brain, 5 f.
vertebrates. *See* chordates
vestibular system, 102
vision, 62–65

visual fields, 65
vocabulary, 184
vocal tract, 22
voice onset time, 109–14, 179
    dipole model, 112, 113 f.
    Spanish v. Chinese, 114
voicing, 109
volley (action potential), 44
Von Baer, K. E., 31
von der Malsburg, C., 72
Von Neumann limit, 14, 74
Von Neumann, J., 14, 43
VOT. *See* voice onset time
vowels, cardinal, 101

walking, 184
Wang, P. P., 176
Washoe (chimpanzee), 22
Watson, J. B., 8, 24
Weaver, W., 166
Wells, W. C., 21
Wernicke, C., 7, 58, 66
Wernicke's area, 62, 66–67, 200 n. 12.1
Wernicke-Lichtheim model of aphasia, 66 f.
West, J. R., 175
Wexler, K., 178
whales, 52
white matter. *See* myelin
Wiesel, T. N., 71, 84, 173, 182
Williams syndrome, 169, 176, 181
word planning, 135
writing, 192
wug test, 141, 177, 200 n. 12.6

XOR (logical operation), 86, 169, 197 n. 5.7